Laboratory Investigations
for Biology

Second Edition

Jean Dickey
Clemson University

Benjamin
Cummings

San Francisco • Boston • New York
Cape Town • Hong Kong • London • Madrid • Mexico City
Montreal • Munich • Paris • Singapore • Sydney • Tokyo • Toronto

Executive Editor : Beth Wilbur
Senior Project Editor: Ginnie Simione Jutson
Marketing Manager: Josh Frost
Production Editor: Steven Anderson
Production Service: Matrix Productions
Copyeditor: Anna Trabucco
Proofreader: Mabel Bilodeau
Text Designer: Gary Head
Cover Designer: Yvo Riezebos
Cover Photograph: Getty/Art Wolfe
Photo Image Manager: Travis Amos
Illustrations: Judy and John Waller, Karl Miyajima, Mark Konrad
Manufacturing Buyer: Vivian McDougal
Compositor and Prepress: Atlis Graphics & Design
Cover Printer: Courier North Chelmsford
Printer: Courier North Chelmsford

ISBN 0-8053-6789-6 (student edition)
ISBN 0-8053-6792-6 (instructor's edition)

Benjamin
Cummings

20 21 V092 16 15

www.aw.com/bc

Contents

15 Circulation   15-1

16 The Sensory System   16-1

Reviewers of the Second Edition

Richard R. Jurin, *University of Northern Colorado*
Craig Laufer, *Hood College*
Elisabeth Martin, *Black Hawk College*
Wilma Patterson, *South Mountain College*
J. Michael Reynolds, *Los Angeles Mission College*

Reviewers of the First Edition

Harold Adams, *Dabbet S. Lancaster Community College*
W. Sylvester Allred, *Northern Arizona University*
Jane Aloi, *Saddleback College*
George Boone, *Susquehanna University*
Jerry Button, *Portland Community College*
Karen Campbell, *Albright College*
Roger Christianson, *Southern Oregon State College*
Robert Creek, *Eastern Kentucky University*
Jean DeSaix, *University of North Carolina*
Terrance Farrell, *Stetson University*
Dino Fiabane, *Community College of Philadelphia*
Kathleen Fisher, *San Diego State University*
Robert Howe, *Suffolk University*
Alan Jaworski, *University of Georgia*
Ted Johnson, *St. Olaf College*
Lee Ann Kirkpatrick, *Glendale Community College*
Mary Rose Lamb, *University of Puget Sound*
Richard Liebaert, *Linn-Benton Community College*
Ivo Lindauer, *University of Northern Colorado*
Raymond Lynn, *Utah State University*
Presley Martin, *Drexel University*
William McComas, *University of Iowa*
Steven McCullough, *Kennesaw State College*
Richard Mortensen, *Albion College*
Paula Piehl, *Potomac State College of West Virginia*
Brian Reeder, *Morehead State University*
Annette Schaefer, *New York City Technical College, City University of New York*
Thomas Shellberg, *Henry Ford Community College*
Guy Steucek, *Millersville University of Pennsylvania*

Preface

All laboratory learning emphasizes activities: Hands-on experiences illustrate concepts or organisms presented in lecture, extend coverage of a topic beyond the textbook, and engage students in the process of scientific inquiry. But "hands-on" should be accompanied by "minds-on." Whether labs are observational, traditional exercises with a known outcome, or student-planned investigations, we can use them to teach students to think as well as do. This laboratory manual emphasizes science process skills as well as content. A unique feature of this manual is that student teams design and perform their own investigations for 8 of the 20 lab topics.

It's going well, but we'll need some plutonium to complete the experiment.

Is this your vision of investigative labs? Does the thought of having introductory students design and perform their own experiments conjure up some unpleasant scenarios? True, introductory students frequently have few laboratory skills and may lack ideas for possible experiments. Prep requirements can be unpredictable too. It may never be quite as easy to run open-ended investigative labs as it is to run traditional or observational labs.

But this Lab Investigations manual provides you with a realistic, practical approach to investigative labs that can be implemented even in large-enrollment courses. This manual and its supplements provide you with the instructional support necessary to meet all the teaching challenges investigative labs may pose.

Why should I use an investigative approach?

As lab instructors we've all had students who have no understanding of what they're doing; they just want to be told what to do next. They plow through one procedure after another, filling in data tables and answer blanks, always seeking the "right" answer, the fact to be memorized later. And though we tell them that there is no "right" answer to a scientific experiment, we ultimately reveal what they "should have gotten" if their data don't match the expected outcome. Thus students can mistakenly learn that science (or the instructor) always has the answer. Science equals facts.

Instead, laboratory learning can teach students to get from one fact to the next by knowing how to read a map rather than by mindlessly following directions. Certainly there are correct answers to many questions that are posed by laboratory material, and there are expected outcomes for most traditional experiments. But whether traditional or investigative, all labs can be designed to offer students the opportunity to think instead of just do, to arrive at answers by integrating background information with observations and results, rather than have answers dropped in their laps.

Investigative labs simply take this idea one step further. In these labs, students can actively participate in *every* phase of an experiment. Because students design their own experiments, even the instructor may not *know* what the results will be. When students engage in the process of how scientific knowledge is acquired, science gains a new meaning in the present, in their lives. As one of our students said after a semester of investigative labs, "Now I understand why it takes so long for the FDA to approve a new drug."

Does it take special skills to teach investigative labs successfully?

It doesn't take special knowledge to teach investigative labs—many of our lab instructors are first-year graduate students—but it does take an openness to students' ideas and a willingness to be more of a mentor than an authority. You have to think like a scientist and a teacher at the same time. Prepare by understanding the methods thoroughly. During the lab session, ask questions and give advice (but not directions) as needed. Let students know that you're there to help them think through problems, but that you don't know all the answers. The Instructor Notes for the investigative labs offer many tips and additional suggestions on facilitating the lab.

Can my students design experiments?

Many students are uncomfortable with the idea of designing their own experiments; they simply have never been asked to do such a thing before. Once they learn, however, they become increasingly confident. All of the investigative labs are organized to support and guide students towards success.

- Lab Topic 1 lays the foundation for student-designed investigations by teaching the components of an experiment.

- Each investigative lab topic begins with a brief demonstration experiment that teaches students how to use one or two methods. (See Lab Topic 6 for an example.)

- The introduction and demonstration experiment for each lab suggest possible independent variables.

- The Instructor Notes at the end of each investigative lab give hints for guiding students through the process of selecting an independent variable and suggest possible investigations.

- All investigative labs include a standard proposal form for students to describe their experimental design (for example, see Exercise 2.4).

- I recommend having students work in teams of three or four for the investigative labs, as working in groups encourages collaborative thought. Teams are often able to solve their own problems, so they rely less on the instructor, and collaboration enables students to perform a more thorough experiment in the small amount of time available.

How can I evaluate students on investigative labs?

Since investigative labs have a different emphasis from traditional labs, grading is also different. Some suggestions for evaluation:

- We collect one proposal from each team and give it a quick-look grade of 10, 8, 6, or 0 points, so grading is not time-consuming.

- After students have performed their experiments, there are two options:

  (a) to have students complete an additional exercise that has them present and interpret their results. (See Exercise 2.5 for an example.) This section is designed in a question-answer format that requires less writing than a lab report and is easier to grade.

  (b) to have students write formal lab reports on their investigations. A Guide to Writing a Scientific Report is included as an appendix to this manual. This guide consists of a sample lab report and a checklist of elements. The checklist can be used to help standardize grading in multiple-instructor courses. You can assign a point value or range of points to each item. With this system, students know in advance how their papers will be evaluated, and instructors are given uniform grading criteria to follow.

Whichever format is used, it is intended that students formalize and interpret their data outside of lab time.

Is there enough time in a 2-hour period to do investigations?

Although the labs are designed for a 3-hour lab period, the Instructor Notes for all the labs suggest how to fit the lab topic into a 2-hour lab period. The 2-hour labs inevitably require some reduction of content and choice, but investigative labs can be done.

Optional schedule. If you want to increase the emphasis on process and communication skills in your lab program, have students do fewer investigations and spend more time on the experiments they do. For example, after students design their experiments, have each team give an oral presentation to the class. With minimal instructor participation, class members critique each other's proposals. The cooperative goal is to have good experiments performed by all teams. Thus students participate in designing several experiments rather than just their own team's. Since this requires extra time, the experiments are not performed until the following week. This schedule works well for a 2-hour period. It also gives the prep staff extra time to anticipate students' needs. After students have obtained results, they can give oral presentations to their classmates, followed by discussion.

Won't this drive the prep staff crazy?

The prep for the investigative lab topics is reasonably predictable. There is a limited number of methods available to the students, and prep directions for each one are given in the Prep Guide, which is free to adopters of this manual. The Prep Guide and the Instructor Notes in this manual have suggestions for additional materials that should be available for students' investigations. See Lab Topics 2, 5, 6, 7, 14, 15, 19, and 20 for examples.

Tips and techniques for accommodating investigative lab prep to your staff and facilities are noted throughout the text.

The Lab Exercises

This lab manual contains 20 labs on a range of topics that are typically covered in an introductory course. The lab topics are classified as investigative, traditional, or observational, but keep in mind that the differences among the categories depend as much on the types of thinking that students are asked to do as on the activities themselves.

Organization

Open-ended investigations are featured prominently in this manual. Eight of the 20 lab topics have students perform introductory exercises and then design and perform their own experiments using scientific process skills.

Traditional investigations. The elements of each experiment are predetermined by the step-by-step procedure in this type of laboratory. But where appropriate, students are asked to identify components of the experiments they perform and make predictions about the results. Most of the emphasis in these lab topics is on interpretation of the results of procedures.

Observational labs stress the theme of adaptation as students observe similarities and differences of taxonomic groups or structures within an organism.

Each of these labs is supported by:

- **Objectives** presented at the beginning of each exercise, which help students focus on the important information.

- **Key terms** printed in bold type so students can locate an explanation of any significant term used in the lab topic.

- **Background information** at the beginning of each lab topic and each exercise that give students a clear idea of the purpose of each activity they perform.

- **Questions** at the end of each activity that ask students to reflect on what they have done. A set of Questions for Review is included at the end of each lab topic.

New in the Second Edition

The second edition of Laboratory Investigations includes two new lab topics. Lab Topic 8, Chromosomes and Cell Division, covers mitosis and meiosis. In addition to pop bead simulations of these processes, students extract DNA so they can actually see the genetic material. Lab Topic 10, Forensic Application of Molecular Genetics, introduces students to the science of DNA profiling by using paper models. Its concluding exercise requires students to apply their knowledge of STR (short tandem repeat) analysis, probability, and Mendelian genetics to solve a mystery. In addition, several lab topics have been updated with new information and activities.

Special Features

Safety Considerations: Students are cautioned about safe use of equipment and hazardous substances by an

icon for general safety

and one for biohazards.

A disposal icon alerts students to use special waste containers for substances that should not be poured down the drain. Safety concerns are also pointed out in the Instructor Notes.

Keys to Success: Special notes provide tips for obtaining good results with the procedures and other reminders.

Color Insert: Eight pages of color photographs illustrate organisms discussed in the labs, especially Plant Diversity and Animal Diversity. All the photographs are cross-referenced to the text. The insert also includes color comparison charts that can be used in place of spectrophotometers for the color change reactions in Lab Topic 5 (Enzymes) and Lab Topic 6 (Cellular Respiration).

Appendices: Appendix A reviews basic measurement techniques and data presentation and interpretation. It can be used as a laboratory exercise, or for student reference. Appendix B, "Guide to Writing Scientific Reports," describes the purpose and content of each section of a formal lab report and illustrates each section with a sample student paper. It also addresses the problem of plagiarism from published sources.

Instructional Support

AIE. The Annotated Instructor's Edition is a source of practical assistance for every type of lab instructor, from first-time teaching assistant to experienced faculty member.

Features of the AIE include:

- Margin Notes
    - (a) Instructor overview. The first page of each lab topic has a thumbnail sketch of its themes and activities.
    - (b) Time requirements for each exercise.
    - (c) Additional information, helpful hints, troubleshooting advice, and suggestions for extending or supplementing the described activities.

- Answers to student questions.

- Sample data and sketches to help prepare first-time instructors for student results.

- Instructor notes at the end of each lab topic discuss organization and presentation of the lab, including suggested content for the introduction and summary. The notes also contain further information and explanations, and suggestions for guiding students through investigative labs. More estimates of time information are also included in the notes.

- Investigative extensions: Each investigative lab topic includes a section that describes possible student investigations. In addition, several other lab topics include suggestions for instructors who want to use an investigative approach in more lab topics, to extend coverage of a topic, or to assign independent projects. See Lab Topics 3, 8, 11, 16, 17, and 18.

- Suggestions for a 2-hour lab. Because the lab topics are written to fill a 3-hour lab period, the Instructor Notes include suggestions for adapting the labs to a 2-hour period.

Supplements

Prep Guide: A thorough Preparation Guide for Laboratory Investigations in Biology accompanies the Lab Investigations manual. It provides:

- Descriptions of the materials used for each exercise, including sources, catalog numbers, recipes, and instructions.

- Suggestions for less expensive alternative materials wherever possible.

- A summary list of the materials required for each lab topic.

- A Planning Guide listing which labs require specific pieces of equipment, living organisms, and solutions.

I hope that you find this lab manual useful and easy to use. If I can provide further assistance, please e-mail, write, or call. I'd also like to hear about your experiences using these lab topics.

Jean Dickey
Biology Instruction
307 Long Hall, Clemson University
Clemson SC 29634–0325
(864) 656-3827
E-mail: dickeyj@clemson.edu

Acknowledgments

I'm grateful for the assistance of many people in the development and preparation of this laboratory manual. Barbara Nicodemus, a talented teacher, made important contributions to both the first and second editions. I've had the benefit of feedback from a great many teaching assistants and undergraduate students at Clemson University. At Clemson, I have especially appreciated the help and support of Darrell Bayles, a former graduate student, and colleagues John Cummings and Bob Kosinski, in developing the investigative lab program. From the start, Benjamin Cummings has provided excellent editorial and production teams. For the second edition, I am very grateful to have had Ginnie Simione Jutson leading the project. With a tight schedule, it was a pretty neat trick for her to keep me on track without making me feel overwhelmed. My thanks also go to Merrill Peterson at Matrix Productions, to Travis Amos, Steve Anderson, and Evelyn Dahlgren at Benjamin Cummings, to Judy and John Waller for the new artwork, and to Anna Trabucco for her copyediting. I remain indebted to former editors Edith Beard Brady, Lisa Donohoe, Korinna Sodic, and Kim Johnson for their dedication to making the concept of an investigative laboratory manual into a reality, and to Neil Campbell for his continuing support.

The Process of Scientific Inquiry

✳ Before coming to lab, you should read through all of Lab Topic 1.

Introduction

Scientific inquiry is a particular way of answering questions. It can't be used for all types of questions. The questions that can be answered by science must meet specific guidelines and scientific investigations must be carried out using certain rules. When an investigation is designed properly and meets these guidelines, then the results are acceptable to other scientists and are added to the body of scientific knowledge. If an investigator cannot show that his or her experiment was done according to the guidelines, then the results of that experiment will not be recognized as valid by other scientists.

The purpose of such guidelines can be understood by comparing them to sports records. For example, a new record set in a track and field event only counts if the meet was approved by the governing body that sets the guidelines. The site and equipment used are scrutinized to be sure that they are within the regulations and the athlete is tested for use of illicit substances. Only when these required conditions are met is the record certified as valid.

In this laboratory you will learn about the basic elements of scientific inquiry and how to apply this process to solving problems.

Outline

Exercise 1.1: The Black Box
Exercise 1.2: Defining a Problem
Exercise 1.3: The Elements of an Experiment
Exercise 1.4: Designing an Experiment
Exercise 1.5: Application of Scientific Inquiry

EXERCISE 1.1
The Black Box

Objective

After completing this exercise, you should be able to

1. Explain the scientific inquiry method, which you apply to various examples in this exercise.

You will use the "black box" exercise as a model of how scientific inquiry is carried out. Each lab team has a container with one or more objects sealed inside. Each team also has an empty container of the same type and a plastic bag holding objects that might be inside the sealed container. Your task is to devise a way to find out what is in the box without opening it. The steps listed below give you some idea of how to proceed. Answer the questions to keep a record of what you did.

Procedure

1. Make observations.

   Investigate the container by any means available to you *except* opening the container.

   What are your observations?

   How did you make your observations?

   What other methods that are not available to you right now might be used to make observations?

   Why is making observations an important first step in solving this problem?

2. Make a guess about the contents of the box.

What did you base your guess on?

3. For now, you still can't open the sealed container. How can you test whether your guess is correct?

4. Use the method you described in step 3 to check your guess. Record your results below. Was your guess correct? How sure are you?

5. If you aren't sure you know yet what is in the box, what should you do next?

6. Short of opening the box, what's the best you can do to find out what's in it?

7. Suppose you tell your instructor what you have concluded is in the box, and he or she says that you are wrong. What are some things that could have led you to make the wrong conclusion?

8. Summarize the methods you used to solve the problem of the black box.

The steps you used to determine the contents of the black box are similar to the procedure followed in one type of scientific investigation. The investigator poses a question—for example, "What is in the box?" From the question and preliminary observations, the investigator makes an educated guess (known as a hypothesis) about the answer. She then devises an experiment to test the hypothesis, performs the experiment, and draws a conclusion from its results. The hypothesis may be revised, and further experiments may be done if the results are not conclusive. Eventually the investigator reaches a point where she is confident that her conclusions are correct.

In Exercises 1.2 and 1.3 you will learn to recognize the elements of a good scientific investigation. In later laboratories you will design your own investigations.

EXERCISE 1.2
Defining a Problem

Objectives

After completing this exercise, you should be able to

1. Identify questions that can be answered through scientific inquiry and explain what characterizes a good question.

2. Identify usable hypotheses and explain what characterizes a good scientific hypothesis.

Every scientific investigation begins with the question that the scientist wants to answer. The questions addressed by scientific inquiry are based on observations or on information gained through previous research, or on a combination of both. Just because a question can be answered doesn't mean that it can be answered *scientifically*. Discuss the following questions with your lab team and decide which of them you think can be answered by scientific inquiry.

What is in the black box?

Are serial killers evil by nature?

What is the cause of AIDS?

Why is the grass green?

What is the best recipe for chocolate chip cookies?

When will the Big Earthquake hit San Francisco?

How can the maximum yield be obtained from a peanut field?

Does watching television cause children to have shorter attention spans?

How did you decide what questions can be answered scientifically?

A scientific question is usually phrased more formally as a **hypothesis,** which is simply a statement of the scientist's educated guess at the answer to the question.

A hypothesis is usable only if the question can be answered "no." If it can be answered "no," then the hypothesis can be proven false. The nature of science is such that we can prove a hypothesis false by presenting evidence from an investigation that does not support the hypothesis. But we cannot prove a hypothesis true. We can only support the hypothesis with evidence from *this particular investigation.* For example, you used hypotheses to investigate the contents of your sealed box. A reasonable hypothesis might have been, "The sealed box contains a penny and a thumbtack." This hypothesis could be proven false by doing an experiment: putting a penny and a thumbtack in a similar box and comparing the rattle it makes to the rattle of the sealed box. If the objects in the experimental box do not sound like the ones in the sealed box, then the hypothesis is proven false by the results of the experiment, and you would move on to a new hypothesis. However, if the two boxes do sound alike, then this does not prove that the sealed box actually contains a penny and a thumbtack. Rather, this investigation has supplied a piece of evidence in support of the hypothesis.

You could test almost any hypothesis you made by putting objects in the empty box. What one hypothesis could *not* be proven false by experimentation?

You may now open the sealed container. Was your final conclusion about its contents correct?

If your conclusion has now been disproven, explain how you reached an erroneous conclusion. (You may have found that your conclusion was wrong in spite of accurate observations and careful experimentation. Conclusions reflect the best evidence available at the time.)

Can you think of any areas of scientific inquiry where a new technology or technique might challenge or disprove hypotheses that are already supported by experimental evidence?

The scientific method applies only to hypotheses that can be proven false through experimentation (or through observation and comparison, a different means of hypothesis testing). It is essential to understand this in order to understand what is and is not possible to learn through science. Consider, for example, this hypothesis: More people behave immorally when there is a full moon than at any other time of the month. The phase of the moon is certainly a well-defined and measurable factor, but morality is not scientifically measurable. Thus there is no experiment that can be performed to test the hypothesis. Propose a testable hypothesis for human behavior during a full moon.

Which of the following would be useful as scientific hypotheses? Give the reason for each answer.

Plants absorb water through their leaves as well as through their roots.

Mice require calcium for developing strong bones.

Dogs are happy when you feed them steak.

An active volcano can be prevented from erupting by throwing a virgin into it during each full moon.

The higher the intelligence of an animal, the more easily it can be domesticated.

The earth was created by an all-powerful being.

HIV (human immunodeficiency virus) can be transmitted by cat fleas.

EXERCISE 1.3
The Elements of an Experiment

Objectives

After completing this exercise, you should be able to

1. Define and give examples of dependent, independent, and standardized variables.
2. Identify the variables in an experiment.
3. Explain what control treatments are and why they are used.
4. Explain what replication is and why it is important.

Once a question or hypothesis has been formed, the scientist turns his attention to answering the question (that is, testing the hypothesis) through experimentation. A crucial step in designing an experiment is identifying the variables involved. **Variables** are things that may be expected to change during the course of the experiment. The investigator deliberately changes the **independent variable.** He measures the **dependent variable** to learn the effect of changing the independent variable. To eliminate the effect of anything else that might influence the dependent variable, the investigator tries to keep **standardized variables** constant.

Dependent Variables

The **dependent variable** is what the investigator measures (or counts or records). It is what the investigator thinks will vary during the experiment. For example, she may want to study peanut growth. One possible dependent variable is the height of the peanut plants. Name some other aspects of peanut growth that can be measured.

All of these aspects of peanut growth can be measured and can be used as dependent variables in an experiment. There are different dependent variables possible for any experiment. The investigator can choose the one she thinks is most important, or she can choose to measure more than one dependent variable.

Independent Variables

The **independent variable** is what the investigator deliberately varies during the experiment. It is chosen because the investigator thinks it will affect the dependent variable. Name some factors that might affect the number of peanuts produced by peanut plants.

In many cases, the investigator does not manipulate the independent variable directly. He collects data and uses the data to evaluate the hypothesis, rather than doing a direct experiment. For example, the hypothesis that more crimes are committed during a full moon can be tested scientifically. The number of crimes committed is the dependent variable and can be measured from police reports. The phase of the moon is the independent variable. The investigator cannot deliberately change the phase of the moon, but can collect data during any phase he chooses.

Although many hypotheses about biological phenomena cannot be tested by direct manipulation of the independent variable, they can be evaluated scientifically by collecting data that could prove the hypothesis false. This is an important method in the study of evolution, where the investigator is attempting to test hypotheses about events of the past.

The investigator can measure as many dependent variables as she thinks are important indicators of peanut growth. By contrast, she must choose only one independent variable to investigate in an experiment. For example, if the scientist wants to investigate the effect that the amount of fertilizer has on peanut growth, she will use different amounts of fertilizer on different plants; the independent variable is the amount of fertilizer. Why is the scientist limited to one independent variable per experiment?

Time is frequently used as the independent variable. The investigator hypothesizes that the dependent variable will change over the course of time. For example, she may want to study the diversity of soil bacteria found during different months of the year. However, the units of time used may be anywhere from seconds to years, depending upon the system being studied.

What was the independent variable in your black box investigation?

What was (or were) the dependent variable(s)?

Identify the dependent and independent variables in the following examples (circle the dependent variable and underline the independent variable):

Height of bean plants is recorded daily for 2 weeks.

Guinea pigs are kept at different temperatures for 6 weeks. Percent weight gain is recorded.

The diversity of algal species is calculated for a coastal area before and after an oil spill.

Light absorption by a pigment is measured for red, blue, green, and yellow light.

Batches of seeds are soaked in salt solutions of different concentrations, and germination is counted for each batch.

An investigator hypothesizes that the adult weight of a dog is higher when it has fewer littermates.

Standardized Variables

A third type of variable is the **standardized variable.** Standardized variables are factors that are kept equal in all treatments, so that any changes in the dependent variable can be attributed to the changes the investigator made in the independent variable.

Since the investigator's purpose is to study the effect of one particular independent variable, she must try to eliminate the possibility that other variables are influencing the outcome. This is accomplished by keeping the other variables at constant levels, in other words, by *standardizing* these variables. For example, if the scientist has chosen the amount of fertilizer as the independent variable, she wants to be sure that there are no differences in the type of fertilizer used. She would use the same formulation and same brand of fertilizer throughout the experiment. What other variables would have to be standardized in this experiment?

Predictions

A hypothesis is a formal, testable statement. The investigator devises an experiment or collects data that could prove the hypothesis false. He should also think through the possible outcomes of the experiment and make **predictions** about the effect of the independent variable on the dependent variable in each situation. This thought process will help him interpret his results. It is useful to think of a prediction as an if/then statement: *If* the hypothesis is supported, *then* the results will be . . .

For example, a scientist has made the following hypothesis: Increasing the amount of fertilizer applied will increase the number of peanuts produced. He has designed an experiment in which different amounts of fertilizer are added to plots of land and the number of peanuts yielded per plot is measured.

What results would be predicted if the hypothesis is supported? (State how the dependent variable will change in relation to the independent variable.)

What results would be predicted if the hypothesis is proven false?

Levels of Treatment

Once the investigator has decided what the independent variable for an experiment should be, he must also determine how to change or vary the independent variable. The values set for the independent variable are called the **levels of treatment.** For example, an experiment measuring the effect of fertilizer on peanut yield has five treatments. In each treatment, peanuts are grown on a 100-m^2 plot of ground, and a different amount of fertilizer is applied to each plot. The levels of treatment in this experiment are set as 200 g, 400 g, 600 g, 800 g, and 1000 g fertilizer/100 m^2.

The investigator's judgment in setting levels of treatment is usually based on prior knowledge of the system. For example, if the purpose of the experiment is to investigate the effect of temperature on weight gain in guinea pigs, the scientist should have enough knowledge of guinea pigs to use appropriate temperatures. Subjecting the animals to extremely high or low temperatures can kill them and no useful data would be obtained. Likewise, the scientist attempting to determine how much fertilizer to apply to peanut fields needs to know something about the amounts typically used so he could vary the treatments around those levels.

Control Treatments

It is also necessary to include **control treatments** in an experiment. A control treatment is a treatment in which the independent variable is either eliminated or is set at a standard value. The results of the control treatment are compared to the results of the experimental treatments. In the fertilizer example, the investigator must be sure that the peanuts don't grow just as well with no fertilizer at all. The control would be a treatment in which no fertilizer is applied. An experiment on the effect of temperature on guinea pigs, however, cannot have a "no temperature" treatment. Instead, the scientist will use a standard temperature as the control and will compare weight gain at other temperatures to weight gain at the standard temperature.

For each of the following examples, tell what an appropriate control treatment would be.

1. An investigator studies the amount of alcohol produced by yeast when it is incubated with different types of sugars. Control treatment:

2. The effect of light intensity on photosynthesis is measured by collecting oxygen produced by a plant. Control treatment:

3. The effect of NutraSweet sweetener on tumor development in laboratory rats is investigated. Control treatment:

4. Subjects are given squares of paper to taste that have been soaked in a bitter-tasting chemical. The investigator records whether each person can taste the chemical. Control treatment:

5. A solution is made up to simulate stomach acid at pH 2. Maalox antacid is added to the solution in small amounts, and the pH is measured after each addition. Control treatment:

Replication

Another essential aspect of experimental design is **replication.** Replicating the experiment means that the scientist repeats the experiment numerous times using exactly the same conditions to see if the results are consistent. Being able to replicate a result increases our confidence in it. However, we shouldn't expect to get exactly the same answer each time, because a certain amount of variation is normal in biological systems. Replicating the experiment lets us see how much variation there is and obtain an average result from different trials.

A concept related to replication is **sample size.** It is risky to draw conclusions based upon too few samples. For instance, suppose a scientist is testing the effects of fertilizer on peanut production. He plants four peanut plants and applies a different amount of fertilizer to each plant. Two of the plants die. Can he conclude that the amounts of fertilizer used on those plants were lethal? What other factors might have affected the results?

When you are designing experiments later in this lab course, consider sample size as an aspect of replication. Since there are no hard and fast rules to follow, seek advice from your instructor regarding the number of samples and the amount of replication that is appropriate for the type of experiment you are doing. Since the time you have to do experiments in lab is limited, inadequate replication may be a weakness of your investigations. Be sure to discuss this when you interpret your results.

Methods

After formulating a hypothesis and selecting the independent and dependent variables, the investigator must find a method to measure the dependent variable; otherwise, there is no experiment. Methods are learned by reading articles published by other scientists and by talking to other scientists who are knowledgeable in the field. For example, a scientist who is testing the effect of fertilizer on peanuts would read about peanut growth and various factors that affect it. She would learn the accepted methods for evaluating peanut yield. She would also read about different types of fertilizers and their composition, their uses on different soil types, and methods of application. The scientist might also get in touch with other scientists who study peanuts and fertilizers and learn about their work. Scientists often do this by attending conferences where other scientists present results of investigations they have completed.

In this course, methods are described in the lab manual or may be learned from your instructor.

Summary

Figure 1.1 summarizes the process of scientific investigation. The process begins and ends with the knowledge base, or what is already known. When a scientist chooses a new question to work on, he first searches the existing knowledge base to find out what information has already been published. Familiarity with the results of previous experiments as well as with the topic in general is essential for formulating a good hypothesis. After working through the rest of the process, the scientist contributes his own conclusions to the knowledge base by presentations at professional meetings and publication in scientific journals. Because each new experiment is built upon past results, the foundation of knowledge grows increasingly solid.

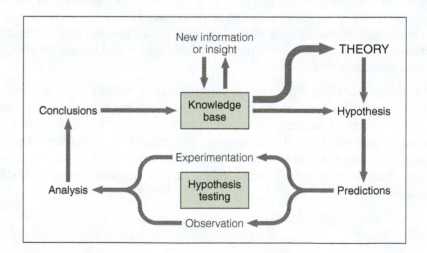

Figure 1.1.
Summary of scientific inquiry.

Scientific knowledge is thus an accumulation of evidence in support of hypotheses; it is not to be regarded as absolute truth. Hypotheses are accepted only on a trial basis. When you read about current scientific studies in the newspaper, keep in mind that the purpose of the media is to report news. In science, "news" is often preliminary results that are therefore quite tentative in nature. It is not unusual to hear that the results of one study contradict another. Some results will hold up under future scrutiny and some will not. However, this does not mean that scientific knowledge is flimsy and unreliable. All scientific knowledge falls somewhere along a continuum from tentative to certain, depending on the evidence that has been amassed. For example, it takes an average of 12 years to get a new drug approved by the FDA as researchers progress through laboratory evaluation of possible compounds, animal studies, and an escalating series of trials in humans. Even so, there are cases of drugs being recalled when new information is discovered. In a way, every medicine you take is still being tested—on you. We don't object to this because we feel confident that the knowledge base is firm, that the science is "done" to an acceptable degree.

EXERCISE 1.4
Designing an Experiment

Objectives

After completing this exercise, you should be able to

1. Given a proposed experiment, critique the experimental design.
2. Given a method for measuring a dependent variable, design an experiment.

Science is almost always a collaborative effort. Teams of scientists work together to solve problems; often these teams include scientists-in-training (graduate students). Working with others brings a variety of perspectives, knowledge bases, and experiences to bear on the problem. When scientists propose a project, they may seek funding from an agency such as the National Science Foundation. In this process, the team's proposal is reviewed by other scientists, who decide whether the problem is worth addressing, whether the proposers have the knowledge required to address it, and whether the design of the experiments is scientifically sound. When scientists have finished an investigation, they present the results to other scientists in the field by making oral presentations at conferences and by publishing articles in scientific journals.

Whether you are going to be a scientist or not, you will find many of the skills that scientists use are applicable to your own career. Most jobs require cooperative effort of some kind, just as you will collaborate with your lab team. Effective communication skills are especially important. You will almost certainly have occasion to present your work or defend your ideas to your coworkers and supervisor.

When you are asked to design an experiment in this course, your lab team will be provided with possible dependent variables and methods and procedures that you can use to measure them. You will decide what independent variable might affect these results. For example, if the topic being studied is the circulatory system, pulse rate and blood pressure might be the dependent variables that you measure to assess cardiovascular fitness. Your lab team would decide what factors (the independent variables) might affect a person's pulse rate and/or blood pressure and then design an experiment to test the effect of one of these factors.

Before you design your own experiments in later laboratories, you will work with your lab team in this part of the laboratory to critique a proposed experiment. This exercise is a rough draft of a proposal that could be improved. Use your knowledge of the scientific process to revise this experiment.

Sample Proposal

Hypothesis

Athletes have better cardiovascular fitness than nonathletes.

Dependent Variable(s)

Pulse rate, blood pressure.

Independent Variable

Athletic training.

Control(s)

Subjects who have had no athletic training (to have a comparison group of subjects); readings taken before exercise (to get a baseline measurement for each subject).

Replication

Three subjects will be used in each group. Each subject will perform the exercise once.

Brief Explanation of Experiment

The pulse rate and blood pressure of athletes and nonathletes will be measured. The subjects will then perform 5 minutes of aerobic exercise. The pulse rate and blood pressure of each subject will be measured again immediately after exercise.

Predictions

We think that the pulse rates and blood pressure of athletes will be lower after exercise and will return to normal rates more quickly than those of nonathletes, indicating better cardiovascular fitness of athletes.

Method

1. Recruit three athletes to be subjects. Our lab team will be the nonathlete subjects.

2. Record resting pulse rate and blood pressure for each subject.

3. All subjects will run up and down the stairs for 5 minutes.

4. Pulse rate and blood pressure of each subject will be measured immediately after the exercise.

5. Pulse rate will continue to be taken until it returns to the resting value. The time taken for each subject's pulse to return to normal will be recorded.

6. Measure and record blood pressure for each person when the resting pulse rate is reached.

Tables Designed to Collect Data

	Resting blood pressure	Resting pulse rate
Athlete 1		
Athlete 2		
Athlete 3		
Tom		
Jennifer		
Eric		

	Blood pressure after exercise	Pulse rate after exercise
Athlete 1		
Athlete 2		
Athlete 3		
Tom		
Jennifer		
Eric		

	Time to reach resting pulse	Blood pressure at resting pulse
Athlete 1		
Athlete 2		
Athlete 3		
Tom		
Jennifer		
Eric		

List other independent variables that might be investigated using these techniques of measuring pulse and blood pressure.

E X E R C I S E   1 . 5

Application of Scientific Inquiry

Objective

After completing this exercise, you should be able to

1. Evaluate research reported in the media using the criteria provided in this exercise.

If you are not a science major, you may be asking why you should learn how scientific inquiry works. One reason is that understanding the science process gives you perspective on the scientific knowledge described in your textbook. When you read a science textbook you are only seeing the tip of the iceberg. Beneath all of that information lies a vast foundation of observation, experimentation, and analysis. Each experiment that you do in lab should give you a tiny taste of how the information in your textbook came to be there—how we know what we know.

Understanding scientific inquiry matters in a personal sense, too. You may never be a producer of scientific knowledge—a scientist—but you make decisions daily as a consumer of science. Do the vitamins and minerals added to Cap'n Crunch cereal make it a health food? Will zinc lozenges shorten my cold? Should I have the Big Mac or the grilled chicken salad? Do I want fries with that? Should I walk to class or drive my SUV? Does it matter whether I get more than 5 hours of sleep a night? Should I steer clear of genetically engineered food? Is a soft drink that has ginseng or ginkgo biloba in it worth the extra cost? Understanding the state of the research can help you make informed choices about issues that affect your health, your wallet, and your impact on our environment. In this exercise, you will learn how to evaluate the science supporting matters that affect you personally.

From Exercises 1.3 and 1.4 you know how to evaluate whether an experiment is designed properly. But accounts of scientific studies in magazines and newspapers don't always provide those details. Some other factors to look for include the following:

Where was the study carried out and who funded it? For objectivity, look for studies funded by competitive agencies such as NIH (National Institutes of Health) and NSF (National Science Foundation). These government-funded organizations require peer review of proposed experiments. That is, a group of knowledgeable scientists determines funding based on criteria such as experimental design and the qualifications of the investigators to perform the experiments. On the other hand, some funding groups have a vested interest in the outcome of research.

How large was the sample size? As you have learned, conclusions drawn from a large sample are more reliable than conclusions from a small sample. Small studies are not bad science, their results are simply more preliminary than those from large studies.

Where was the study published? The most reliable sources are peer-reviewed journals. Knowledgeable scientists critique each article submitted to determine whether it merits publication. Other types of publications do not require the same rigorous screening. For example, an industry newsletter or a Web site promoting herbal supplements will probably not offer a thorough and objective review of the research.

Procedure

Your instructor may have asked you to find a news article yourself, or she may have one to assign to you. In either case, read the article and briefly describe the following aspects of the science reported on. (Your article may not include all of these.)

Experimental Design:

Independent variable —

Dependent variable(s) —

Standardized variables —

Control treatment —

How large was the sample size?

Where was the study carried out, and who funded it?

Where was the study published?

How convinced are you by the results reported in the article? Rate your conclusion on a scale of 0 (this isn't even science!) to 3 (I would base a personal decision on these results). Explain how you arrived at your rating.

Questions for Review

1. A group of students hypothesizes that the amount of alcohol produced in fermentation depends on the amount of glucose supplied to the yeast. They want to use 5%, 10%, 15%, 20%, 25%, and 30% glucose solutions.

   a. What is the independent variable?

   b. What is the dependent variable?

c. What control treatment should be used?

d. What variables should be standardized?

2. Having learned the optimum sugar concentration, the students next decide to investigate whether different strains of yeast ferment glucose to produce different amounts of alcohol. Briefly explain how this experiment would be set up.

3. A group of students wants to study the effect of temperature on bacterial growth. To get the bacteria, they leave petri dishes of nutrient agar open on a shelf. They then put the dishes in different places: an incubator (37°C), a lab room (21°C), a refrigerator (10°C), and a freezer (0°C). Bacterial growth is measured by estimating the percentage of each dish covered by bacteria at the end of a 3-day growth period.

   a. What is the independent variable?

   b. What is the dependent variable?

   c. What variables should be standardized?

4. A team of scientists is testing a new drug, XYZ, on AIDS patients. They expect patients to develop fewer AIDS-related illnesses when given the drug, but they don't expect XYZ to cure AIDS.

   a. What hypothesis are the scientists testing?

   b. What is the independent variable?

   c. What is the dependent variable?

d. What control treatment would be used?

e. What variables should the researchers standardize?

5. A group of students decides to investigate the loss of chlorophyll in autumn leaves. They collect green leaves and leaves that have turned color from sugar maple, sweet gum, beech, and aspen trees. Each leaf is subjected to an analysis to determine how much chlorophyll is present.

   a. What is a reasonable hypothesis for these students?

   b. What is the independent variable?

   c. What is the dependent variable?

   d. What would you advise the students about replication for this experiment?

6. A scientist wants to study mating behavior in crickets. She hypothesizes that males that win the most male-vs.-male contests mate with the most females. She observes the crickets to obtain data. For each male, she counts the number of male-male fights he wins and the number of females he mates with.

   a. What is the independent variable?

   b. What is the dependent variable?

   c. What constitutes replication in this experiment?

pH and Buffers

Introduction

Molecules that are dissolved in water may separate (ionize) into charged fragments. pH is a measure of the concentration of one of those charged fragments, hydrogen ions (H^+), in solution. A substance that has a high concentration of H^+ is acidic. A substance that has a low concentration of H^+ is basic (alkaline).

The pH scale ranges from 0 (most acidic) to 14 (most basic). There is a tenfold difference between pH units. For example, a solution with a pH value of 6 has a ten-times-greater concentration of hydrogen ions than a solution with a pH value of 7. Some examples are shown in Table 2.1.

Outline

Exercise 2.1: Introduction to Acids, Bases, and pH
Exercise 2.2: Using Red Cabbage Indicator to Measure pH
    Activity A: Making a Set of Standards
    Activity B: Comparing pH of Beverages and Stomach Medicines
Exercise 2.3: Using the pH Meter to Determine Buffering Capacity
Exercise 2.4: Designing an Experiment
Exercise 2.5: Performing the Experiment and Interpreting the Results

---

EXERCISE 2.1
Introduction to Acids, Bases, and pH

Objectives

After completing this introductory exercise, you should be able to

1. Explain what makes a solution acidic or basic.
2. Explain the pH scale.
3. Describe the phenol red test for pH.

An **acid** is a substance that releases or causes the release of H^+ into solution. Solutions that have pH values lower than 7 are considered to be acids. Some common acids are hydrochloric acid, acetic acid, carbonic acid, and sulfuric acid. All of these compounds contain hydrogen. When

Table 2.1
pH Scale

pH	Relative strength	Examples
0	Strong acid	
1		Battery acid
2		Gastric fluid
3	Moderate acid	Orange juice
4		Tomato juice
5	Weak acid	
6		Rainwater Milk
7	Neutral	Pure water Blood
8		
9	Weak base	Baking soda Milk of Magnesia
10		
11	Moderate base	Household ammonia
12		
13	Strong base	Hair remover
14		Oven cleaner

the compound is dissolved in water, hydrogen ions are released, and the pH of the solution is low. Your instructor will use an indicator called **phenol red** to demonstrate the acidity of hydrochloric acid. Phenol red is red when the solution is basic and turns yellow in acidic solution.

What color is the phenol red solution initially?

What happens when hydrochloric acid is added?

A compound does not have to contain hydrogen ions itself in order to be an acid. Carbon dioxide (CO_2), for example, can combine with water to generate H^+. Your instructor will use phenol red again to demonstrate that CO_2 is an acid. Briefly describe this demonstration.

The reaction of SO_2 (sulfur dioxide) is similar to that of CO_2. The presence of SO_2 in the atmosphere is partially responsible for acid rain.

A **base** is a substance that can remove H^+ from solution, thus lowering the concentration of H^+. Many bases ionize to produce hydroxyl ions (OH^-), which combine with H^+ to make water (H_2O). Some common bases are sodium hydroxide ($NaOH$), magnesium hydroxide ($Mg(OH)_2$), and potassium hydroxide (KOH).

Mixing an acid with a base can produce a neutral solution by combining the H^+ with the OH^- to make water (H_2O). Pure water, which ionizes to produce equal numbers of H^+ and OH^-, is neutral (pH 7). Don't expect to get a pH of 7 when you measure water in the lab, though. Tap water contains impurities, and its pH varies a great deal. Distilled water is weakly acidic; its pH is usually around 6.

It is important for organisms to maintain a constant internal pH. As you will learn in later laboratories, biological molecules, especially proteins, are sensitive to pH, and they may not function correctly when the pH is changed.

In the following exercises you will make an indicator solution to measure pH and also learn how to use a pH meter. You will determine the pH values of some common substances and investigate how buffer systems work to maintain a constant pH.

 Wear safety glasses while performing these exercises. Strong acids and strong bases are corrosive. Inform your instructor immediately if any solution is spilled or comes in contact with your skin or clothing.

E X E R C I S E   2 . 2
Using Red Cabbage Indicator to Measure pH

Objectives

After completing this exercise, you should be able to

1. Explain what a pH indicator is used for.
2. Describe how to measure pH using red cabbage indicator.

Several methods are available for determining pH. Many of these methods rely on the ability of certain chemicals called **indicators** to change color, depending on the pH of the surrounding solution. Papers saturated with indicators, such as litmus paper and alkacid test paper, can also be used.

An indicator can easily be made from a solution of anthocyanins, the pigments responsible for red, blue, and purple colors in flowers, fruits, and autumn leaves. These pigments change color as the pH changes. Red cabbage is loaded with anthocyanins, so we can make a pH indicator by boiling red cabbage to extract the pigments. Your instructor will make the extract at the beginning of class.

The use of standards, a set of known quantities, is an important technique in biological research. By comparing unknowns with the standards, we can determine what we want to know about the unknowns. The color of cabbage extract depends on the pH of the solution it is in. Your set of standards will show the color of the cabbage extract at pH 2, 4, 6, 7, 8, 10, and 12. You will then determine the pH values of various substances by mixing each substance with cabbage extract and comparing its color to the standards.

Activity A: Making a Set of Standards

Your lab team should make a set of standards using the cabbage extract and solutions of known pH according to the following procedure.

Procedure

1. Put seven clean test tubes in a rack and label them 2, 4, 6, 7, 8, 10, and 12.
2. Pipet 5 mL of the appropriate buffer into each tube (pH 2 buffer into the tube labeled 2, and so on).
3. Get a dropping bottle of cabbage extract from your instructor. Add 3 mL of cabbage extract to each tube.
4. Cover the tubes with Parafilm and mix well.
5. Record the color in each tube in Table 2.2.

✳ **Save this set of standards for use throughout the lab period.**

Table 2.2
Color of Standard Solutions for
Red Cabbage Indicator

pH	Color
2	
4	
6	
7	
8	
10	
12	

Record both the initial and final colors
at pH 12. The pigments are not stable
at this pH.

Activity B: Comparing pH of Beverages and Stomach Medicines

Look at Table 2.3 to see what aspect of pH is being investigated in this experiment, and answer the following questions.

Table 2.3
pH Values of Beverages and Medicines

Beverages	pH	Medicines	pH
White grape juice		Milk of Magnesia $(Mg(OH)_2)$	
7-Up		Sodium bicarbonate $(NaHCO_3)$	
White wine		Maalox	
Seltzer water			

What hypothesis could be tested with this experiment?

What is the independent variable in this experiment?

What is the dependent variable?

What substance could be used as a control for this experiment?

Predict the outcome of the experiment in terms of your hypothesis. What results will support the hypothesis? What results will prove the hypothesis false?

Use the cabbage indicator method to measure the pH of the substances listed in Table 2.3.

Procedure

1. Put 2 droppersful of the solution to be tested in a clean test tube.
2. Add 1 dropperful of cabbage extract.
3. Swirl the tube gently to mix.
4. Compare the color of the solution to the colors of your cabbage indicator standards.
5. Record the pH value for each substance in Table 2.3.
6. Measure and record the pH of your control.

    Control: _____

    pH of control: _____

Was your hypothesis proven false or supported by the results? Use data to support your answer.

What components of the beverages you tested might be responsible for the pH values of the beverages?

What components of the medicines you tested might be responsible for the pH values of the medicines?

Explain why stomach medicines should have pH values that are much higher than normal stomach pH, which is around 2. (Hint: Why do people take these medicines?)

EXERCISE 2.3
Using the pH Meter to Determine Buffering Capacity

Objectives

After completing this exercise, you should be able to

1. Define buffer, and explain why buffers are important to organisms.
2. Describe how to use a pH meter.
3. Interpret a titration curve (graph of pH versus milliliters of HCl and NaOH added) to determine whether a solution has buffering capacity and, if so, over what pH range.
4. Explain why some solutions have buffering capacity and others don't.

In order for normal physiological processes to occur, pH must remain relatively constant. An excess of H^+ or OH^- can interfere with the functioning of biological molecules, especially proteins. In our bodies, for example, blood pH is usually maintained between 7.3 and 7.5. However, blood returning to the heart contains CO_2 picked up from the tissues, which lowers the blood pH. Metabolic reactions in cells may contribute an excess of hydrogen ions. Our diets may also affect blood pH. Several buffering systems keep the pH constant.

A **buffer** is a solution whose pH resists change on addition of small amounts of either an acid or a base. To be a good buffer, a solution should have a component that acts as a base (takes H^+ out of solution) and a component that acts as an acid (puts more H^+ into solution when there is an excess of OH^-).

The buffering capacity of a solution is tested by adding small amounts of acid (for example, HCl) and base (for example, NaOH) and checking the pH after each addition. If the pH changes only slightly, the solution is a good buffer. Eventually its buffering capacity will be exhausted, however, and the pH will change dramatically.

A buffer operates in a specific pH range. The buffering systems in our blood, for example, buffer at around pH 7.4. That is, they maintain the pH at or very close to 7.4. The solutions you used to make up your standards for red cabbage indicator maintain each buffer at a certain pH. The pH 2 buffer maintains pH at 2, the pH 4 buffer maintains pH at 4, and so on. Notice that the purpose of a buffer is *not* to make the pH neutral (7).

Figure 2.1.
Titration curve for an
unknown solution.

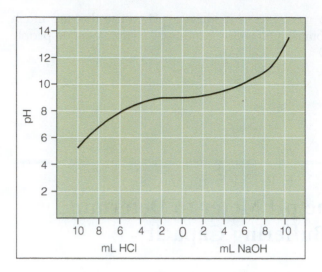

Look at Figure 2.1. Is this solution a good buffer? Explain how you know.

At what pH does the solution buffer?

Figure 2.2.
Titration curve for an unknown
solution.

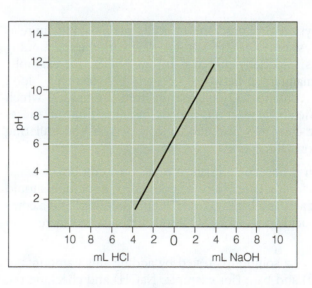

Look at Figure 2.2. Is this solution a good buffer? Explain how you know.

You will use a pH meter to test buffering capacity in this exercise. The pH meter has a sensitive electrode that measures the H^+ concentration in solution. It can measure in tenths of pH units; some models measure hundredths of pH units.

Most of the control knobs are used only to calibrate the machine (a buffer of known pH is used to standardize the pH meter). Figure 2.3 illustrates

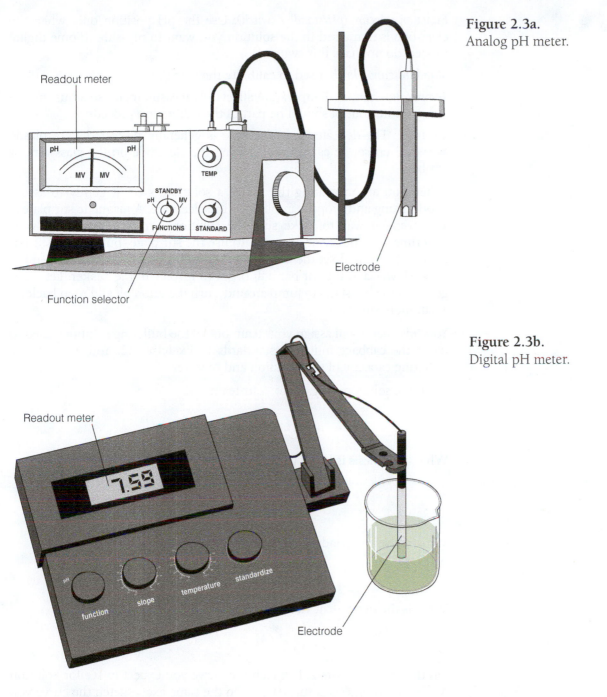

Figure 2.3a.
Analog pH meter.

Figure 2.3b.
Digital pH meter.

two commonly used types of pH meters: digital and analog. Your instructor can help you identify the parts you will need to use for the model of pH meter available in your laboratory.

Familiarize yourself with the pH meter in your laboratory by locating the following parts.

Controls

Readout meter: Shows the pH of the solution. On an analog meter, there are usually two scales. One shows pH and the other shows millivolts. You will use the pH scale. (If you are using a digital pH meter, only the number representing the solution's pH will be displayed.)

Function selector (pH/standby switch): Use the pH position only when the electrode is immersed in the solution you want to measure. (Some digital models do not have this switch.)

Standardization knob: Used to calibrate the machine.

Temperature control: Temperature affects pH measurement, so adjusting the temperature control should be part of the calibration procedure.

Electrode: The delicate glass electrode is generally protected by a plastic sleeve. Even so, be careful not to bang the electrode on the glassware or stir bar.

When you are measuring the pH of a solution, swirl it gently to assure good mixing and proper sampling by the electrode. A magnetic stir plate is a convenient way to make sure the solutions are well mixed for your experiment on buffering capacity. To use the stir plate, put a small stir bar in the beaker, and set the beaker in the middle of the stir plate. Turn the knob slowly, and the stir bar will begin to revolve in the beaker. Let it stir gently. If the bar starts to jump around, turn the knob off and then back on again more slowly.

Your instructor will assign your team one of the buffering solutions used to make the cabbage indicator standards in Exercise 2.2. You will test the buffering capacity of that solution and of water.

Buffering solution assigned to your team:

What hypothesis is being tested?

What is the independent variable in this experiment?

What is the dependent variable?

On the axes of Figure 2.4, sketch the curve you expect to see for Solution X if your hypothesis is supported. On the same axes, sketch the curve you expect to see if your hypothesis is proven false.

Why should you determine the buffering capacity of water as part of this experiment?

Procedure

1. Pour 40 mL of your assigned solution into a 100-mL beaker. Put a stir bar in the beaker and put it on the magnetic stirrer.

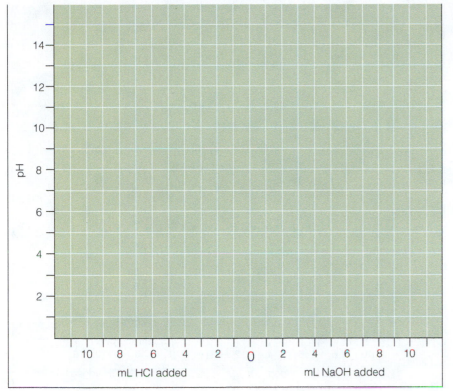

Figure 2.4.
Predicted results of buffering capacity experiment.

2.   Determine the pH of the solution by following steps a–d below.

   a. Raise the electrode out of the soak beaker and rinse it with distilled water from the wash bottle.

   b. Immerse the electrode in the solution you want to measure. Swirl the beaker gently. If you are using a magnetic stir plate, make sure the bar clears the electrode before you turn it on.

   c. If your pH meter has a function switch, change it from standby to pH.

   d. Read the pH value on the readout meter or digital display.

3.   Record the pH value at 0 on the x-axis of Figure 2.5 on the next page.

4.   Add 1.0 mL of 0.1N HCl. N stands for normal, a measure of concentration. (If you're using a magnetic stirrer, you can leave it on with the electrode immersed throughout the procedure. If you must take the electrode out of the beaker to mix, turn the function switch to standby first.)

5.   Record the new pH at "1 mL HCl added" on the x-axis of Figure 2.5.

6.   Add another 1 mL of HCl and record the new pH at "2 mL HCl added" on the x-axis of Figure 2.5.

7.   Continue to add 1 mL of HCl at a time and record the pH until you have added 10 mL *or* there is a significant decrease in pH, whichever comes first.

8.   If the pH meter has a function switch, turn it back to standby.

9.   Raise the electrode out of the solution.

Figure 2.5.
Results of buffering capacity
experiment.

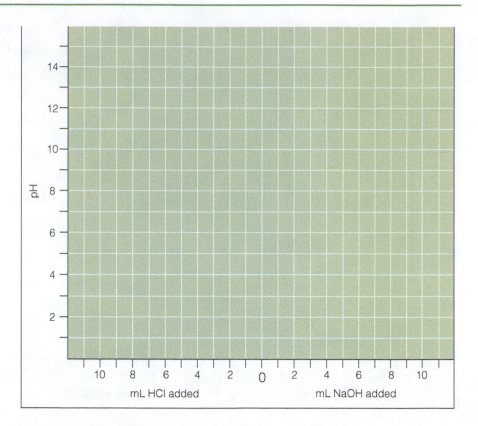

10. Rinse the electrode with distilled water, and wipe it with a cleaning tissue.

 Always rinse the electrode with distilled water after use to avoid contamination of solutions.

11. Dispose of your solution, and rinse and dry the beaker and stir bar.

12. Put another 40 mL of the same solution into the beaker and repeat the procedure using $0.1N$ NaOH instead of HCl until you have added 10 mL *or* there is a significant increase in pH, whichever comes first. Record the pH values on the "mL NaOH added" side of the x-axis of Figure 2.5.

13. When you are finished determining the buffering capacity of the solution you were assigned, repeat steps 1–12 to determine the buffering capacity of water. Use the axes in Figure 2.5 to graph the results.

When you're done with this procedure, leave the electrode immersed in some solution (water or a buffer). It should never be allowed to dry out.

Write a descriptive title for Figure 2.5.

Review your prediction (Figure 2.4) about the buffering capacity of the solution you were assigned. Was your prediction correct?

If the solution is a buffer, at what pH does it buffer?

Evaluate your hypothesis. Is it supported or proven false by the results?

Describe the buffering capacity (or lack of it) of this solution.

Describe the buffering capacity (or lack of it) of water.

E X E R C I S E   2 . 4
Designing an Experiment

Objective

After completing this exercise, you should be able to

1. Design an original experiment to investigate some aspect of pH or buffering capacity.

In Exercises 2.2 and 2.3 you learned a method of measuring pH and a method of determining buffering capacity. In Exercises 2.4 and 2.5, your lab team will design an experiment using one of these methods, perform your experiment, and present and interpret your results.

The following materials will be supplied for your group.

For pH Measurement Using Red Cabbage Indicator

You will already have the standards made up from Exercise 2.2.
extra test tubes
pasteur pipets and rubber bulbs

For Determining Buffering Capacity

You will use the same materials used in Exercise 2.3. Your instructor will be able to tell you what additional materials will be available.

Proposed Experiment

If you are considering an experiment using red cabbage indicator, you might want to review Table 2.1 for ideas. Also, try to think of substances whose function may be pH dependent. For example, you may recall hearing the terms pH, acid, or base used in advertisements. If you are planning an investigation of buffering capacity, consider what substances might be expected to be good buffers.

Describe your experiment below.

Question or Hypothesis

Dependent Variable

Independent Variable

Explain why you think this independent variable will affect pH or buffering capacity.

Control Treatment(s)

Replication

Brief Explanation of Experiment

Predictions (What results would support your hypothesis? What results would prove your hypothesis false?)

Method

Design a Table to Collect Your Data

List Any Additional Materials You Will Require

E X E R C I S E   2 . 5
Performing the Experiment and Interpreting the Results

Objectives

After completing this exercise, you should be able to

1. Perform the experiment your lab team designed.
2. Present and interpret the results of your experiment.

Before you do the experiment, be sure that everyone on your lab team understands the techniques that will be used. You may want to divide up the tasks before you begin work.

Be thorough in collecting data. Don't just write down numbers; record what they mean as well. Don't rely on your memory for information that you need when reporting on your experiment later! If you have any questions, doubts, or problems during the experiment, be sure to write them down, too.

Results

Before you begin to prepare your results for presentation, decide on the best format to use. Remember, you want to give the reader a clear, concise picture of what your experiment showed. Refer to the data presentation section of Appendix A (Tools for Scientific Inquiry) for help. If you are drawing graphs, use graph paper. Complete your tables and/or graphs before attempting to interpret your results.

Write a few sentences *describing* the results (don't explain why you got these results or draw conclusions yet).

Discussion

Look back at the hypothesis or question you posed in this experiment. Look at the graphs or tables of your data. Do your results support your hypothesis or prove it false? Explain your answer, using your data for support.

Did your results correspond to the prediction you made? If not, explain how your results are different from your expectations and why this might have occurred.

Describe how your data are supported by information from other sources (for example, textbooks or other lab teams working on the same problem).

If you had any problems with the procedure or questionable results, explain how they might have influenced your conclusion.

If you had an opportunity to repeat and extend this experiment to make your results more convincing, what would you do?

Summarize the conclusion you have drawn from your results.

Questions for Review

1. You have blown air from your lungs into a solution of phenol red and changed its color from red to yellow. Suggest a way to turn the color back to red.

2. Give an example of two substances that, when mixed together, will produce a neutral solution.

3. You measure the pH of your garden soil and find that it is 6. You measure the pH of peat moss and find that it is 4. How much greater is the concentration of hydrogen ions in peat moss than in the garden soil?

4. What's one ingredient that could make soft drinks acidic?

5. Aspirin has a pH of 3. Some people who take large amounts of aspirin (for example, for arthritis) take a pill that combines aspirin with Maalox. What's the purpose of this combination?

6. If you want to do an experiment to measure buffering capacity, why is red cabbage indicator not a good choice of methods?

7. On the graph in Figure 2.6, which compound is the best buffer? Explain why. Over what pH range does the compound buffer?

Figure 2.6.
Graph of buffering activity of three different compounds.

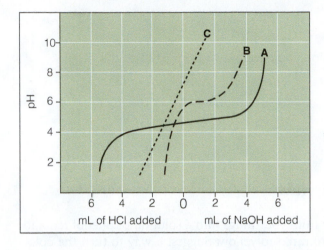

8. Considering that CO_2 is produced as a by-product of cellular metabolism, why is it important for our blood to contain buffers?

Macromolecules

Introduction

Living organisms are composed of molecules that come in diverse shapes and sizes and serve a variety of purposes. Some molecules form the structure of an organism's body—for example, the cellulose that makes up the cell walls in plants, the proteins and phospholipids that comprise cell membranes, and the fibers that make up animal muscles.

There is also a wide array of molecules that perform all the functions of life. For example, enzymes catalyze the chemical reactions necessary for biological processes, neurotransmitters convey information from one brain cell to another, and visual pigments absorb light so that you can read the words on this page.

In this laboratory you will study three classes of the largest biological molecules, called **macromolecules:** carbohydrates, lipids, and proteins. A fourth class of macromolecules, the nucleic acids, will not be studied in this laboratory.

Outline

Exercise 3.1: Carbohydrates

    Activity A: Monosaccharides and Disaccharides

    Activity B: Starch

    Activity C: Hydrolysis of Carbohydrates

Exercise 3.2: Lipids

Exercise 3.3: Proteins

Exercise 3.4: Macromolecules in Food

    Activity A: Separation of Butter

    Activity B: Tests with Food

EXERCISE 3.1
Carbohydrates

Objectives

After completing this exercise, you should be able to

1. Define monosaccharide, disaccharide, and polysaccharide and give examples of each.
2. Name the monosaccharide components of sucrose and starch.

3. Describe the test that indicates the presence of most small sugars.
4. Describe the test that indicates the presence of starch.
5. Define hydrolysis and give an example of the hydrolysis of carbohydrates.

Most **carbohydrates** contain only carbon (C), oxygen (O), and hydrogen (H). The simplest form of carbohydrate molecules are the **monosaccharides** ("single sugars"). One of the most important monosaccharides is glucose ($C_6H_{12}O_6$), the end product of photosynthesis in plants. It is also the molecule that is metabolized to produce another molecule, ATP, whose energy can be used for cellular work. There are many other common monosaccharides, including fructose, galactose, and ribose.

Some **disaccharides** ("double sugars") are also common. A disaccharide is simply two monosaccharides linked together. For example, maltose consists of two glucose molecules, lactose (milk sugar) consists of glucose and galactose, and sucrose (table sugar) consists of glucose and fructose. Can you discern a rule used in naming sugars?

Carbohydrates are also found in the form of **polysaccharides** ("many sugars"), which are long chains of monosaccharide subunits linked together.

Starch, a polysaccharide composed of only glucose subunits, is an especially abundant component of plants. Most of the carbohydrates we eat are derived from plants. What was the last starch you ate?

Starch is the plant's way of storing the glucose it makes during photosynthesis. When you eat starch, you are consuming food reserves that the plant has stored for its own use. The starch of potatoes and root vegetables, for example, would be used the next spring for the plant's renewed growth after the winter die-back. All perennial plants (those that come up year after year, such as tulips) have some kind of food storage for overwintering. Beans, on the other hand, contain starch in the seeds. Beans are annual plants; they will die at the end of the growing season. So the seeds are stocked with starch to use when they have a chance to germinate the next spring.

Animals store glucose in **glycogen,** which is another form of polysaccharide. Although starch and glycogen are both composed of glucose subunits, the glucose molecules are bonded together in different ways, so these polysaccharides are not identical. Glucose subunits are bonded together a third way in the polysaccharide **cellulose.** While starch and glycogen are meant to be metabolized for energy, cellulose, which is the most abundant carbohydrate in the world, is a structural molecule that is designed *not* to be metabolized. Cellulose makes up the cell walls of plants and is a primary component of dietary fiber. For most animals it is completely indigestible. Those that can digest it, such as termites and cows, do so only with the assistance of organisms such as bacteria, fungi, or protistans.

Most disaccharides and polysaccharides can be broken down into their component monosaccharides by a process called **hydrolysis**, which is accomplished in organisms by digestive enzymes. This process is important in seeds. If the seed's food resource is starch, it must be able to convert the starch to glucose. The glucose is then used to generate ATP, which in turn is used to provide the growing plant embryo with energy for metabolic work. Hydrolysis of starch begins when the seed takes up water and begins to germinate.

Germination of barley seeds is part of the process of brewing beer. When the barley is germinated, the starch-to-sugar conversion begins. In the breakdown of starch, disaccharide maltose molecules are formed before the final product, glucose, is obtained. At a certain point in the germination, the barley is dried so that no further hydrolysis takes place. The maltose sugar is extracted and used in the brewing process. That's the "malt" listed on the beer can as an ingredient. The process of germinating the barley is called malting.

A chemical hydrolysis can be done in the laboratory by heating the molecules with acid in the presence of water. You will perform a chemical hydrolysis in this exercise.

 Wear safety glasses throughout the lab session.

Activity A: Monosaccharides and Disaccharides

You will use **Benedict's reagent** as a general test for small sugars (monosaccharides and disaccharides). When this reagent is mixed with a solution containing single or double sugars and then heated, a colored precipitate (solid material) forms. The precipitate may be yellow, green, orange, or red. If no monosaccharide or disaccharide is present, the reaction mixture remains clear. However, Benedict's reagent does not react with all small sugars. For example, sucrose gives a negative Benedict's reaction.

Glucose will be used in this laboratory to demonstrate a positive Benedict's test (Figure 3.1). What should be used as a negative control for this test?

Procedure

1. Make a boiling water bath by filling a beaker about half full of water and heating it on a hot plate. Put six or seven boiling chips in the beaker. You will need to use this water bath in several activities.

 Set the hot plate where it will not be in your way as you work. Be careful—it will be very hot!

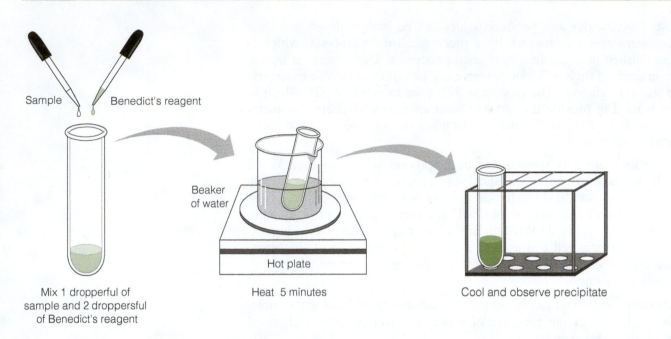

Sample   Benedict's reagent

Beaker of water

Hot plate

Mix 1 dropperful of sample and 2 droppersful of Benedict's reagent

Heat 5 minutes

Cool and observe precipitate

Figure 3.1.
Benedict's test for detecting small sugars.

2. Get two test tubes and label them 1 and 2 with a wax pencil.

✳ **Make heavy marks so that they don't melt off in the water bath.**

3. Put 1 dropperful of glucose into Tube 1. Tube 1 is the positive control.
4. Tube 2 is the negative control. What substance goes in it? How much should be used?

5. Add 2 droppersful of Benedict's reagent to each tube.
6. Place the tubes in the boiling water bath and let them heat for 5 minutes.
7. After 5 minutes, remove the tubes from the water bath.

 Use a test tube holder to retrieve test tubes from the boiling water.

8. Allow the tubes to cool at room temperature for several minutes in the test tube rack while you go on to the next procedure.

9. Record your observations below.

Tube 1 (glucose):

Tube 2 (negative control):

Interpretation of Results

Describe a positive Benedict's test.

What are the limitations of this test?

Activity B: Starch

Starch is tested by using **iodine reagent** (I_2KI—iodine potassium iodide). A dark blue color indicates the presence of starch (Figure 3.2).

You will use a solution of potato starch to demonstrate a positive test. What negative control should be used for this test?

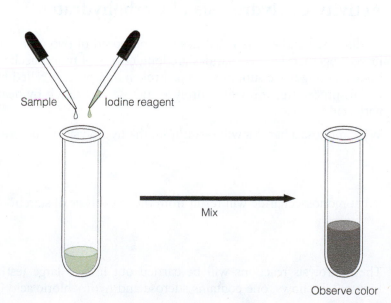

Sample   Iodine reagent

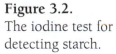

Mix

Observe color

Figure 3.2.
The iodine test for detecting starch.

Procedure

1. Get two test tubes and label them 1 and 2.
2. Put a dropperful of starch solution in Tube 1. This is the positive control.
3. Tube 2 is the negative control. What substance goes in it? How much should be used?

4. Put 3 or 4 drops of iodine reagent into each tube.
5. Record the results below.
    Tube 1 (starch):

    Tube 2 (negative control):

Interpretation of Results

Describe a positive test for starch.

What are the limitations of this test?

Activity C: Hydrolysis of Carbohydrates

As discussed earlier, disaccharides are composed of two monosaccharides linked together. Polysaccharides are long chains of monosaccharides. The bonds joining these subunits can be broken in a process called hydrolysis. In this procedure, you will hydrolyze sucrose and starch by heating them with acid.

What monosaccharides will result from the hydrolysis of sucrose?

What monosaccharide will result from the hydrolysis of starch?

The hydrolysis reactions will be carried out in two large test tubes. As Figure 3.3 shows, one contains sucrose and hydrochloric acid (HCl) and

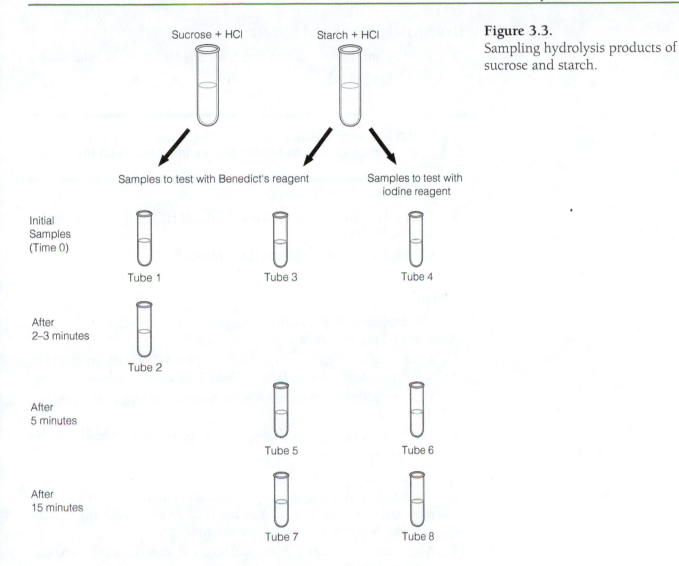

Figure 3.3.
Sampling hydrolysis products of sucrose and starch.

the other contains starch and HCl. You will sample the sucrose tube twice: once before the hydrolysis has begun and again after 3 minutes. You will take six samples from the starch tube: two before the hydrolysis has been done, two after 5 minutes of hydrolysis, and two after 15 minutes. Two samples are needed at each time so that one can be tested for small sugars (Benedict's test) and one can be tested for starch (iodine test).

Procedure

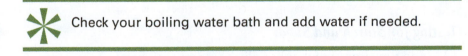

✳ **Check your boiling water bath and add water if needed.**

1. Get eight test tubes and label them 1 through 8. Line up the test tubes in order in a test tube rack.
2. Get two extra large test tubes and label them starch and sucrose. Use an empty beaker as a test tube holder if the test tubes don't fit in the rack.

Hydrolysis of Starch and Sucrose

3. Pipet 10 mL starch solution and 5 mL 2*N* HCl into the tube labeled starch.

> ⚠️ **HCl is a strong acid. Handle it with caution. After you use the pipet, replace it in its holder. Do not lay it down on the bench.**

4. Pipet 10 mL sucrose solution and 2 mL 2*N* HCl into the tube labeled sucrose.

5. Swirl each tube gently to mix the contents.

Sampling

6. Use a pasteur pipet to draw 1 pipetful of solution from the sucrose tube and put it in Tube 1.

7. Using a *different* pasteur pipet, draw 1 pipetful of solution from the starch tube and put it in Tube 3. (Skip Tube 2 for now.)

8. Draw an additional pipetful of solution from the starch tube and put it in Tube 4.

9. Place the extra-large starch and sucrose tubes in your boiling water bath. Note the time:

    _____

10. After 2 or 3 minutes, draw 1 pipetful of solution from the sucrose tube and put it in Tube 2. You are now finished with the sucrose solution. You may remove it from the water bath.

11. After 5 minutes, draw 1 pipetful of solution from the starch tube and put it in Tube 5.

12. Put a second pipetful of starch solution in Tube 6.

13. Wait 10 more minutes and then repeat steps 11 and 12, putting the solution in Tubes 7 and 8.

 **Do Exercise 3.2 during the waiting period.**

Testing for Starch and Sugar

14. Add two droppersful of Benedict's reagent to Tubes 1, 2, 3, 5, and 7. Place these tubes in the boiling water bath for 5 minutes.

15. Add 3 or 4 drops of iodine reagent to Tubes 4, 6, and 8. Record the results in Table 3.1 on the next page.

Table 3.1

	Tube Number							
	Sucrose		Starch					
	1	**2**	**3**	**4**	**5**	**6**	**7**	**8**
Time (min)	0	2–3	0	0	5	5	15	15
Benedict's reagent				▓		▓		▓
Iodine reagent	▓	▓	▓		▓		▓	

16. Remove the tubes from the water bath and wait 5 minutes for them to cool. Record the results in Table 3.1.

Interpretation of Results

Explain the results you obtained using the Benedict's test on the sucrose solution.

Explain the results you obtained using the iodine reagent test with starch.

Explain the results you obtained using the Benedict's test with starch.

Why does hydrolysis of starch take longer than hydrolysis of sucrose?

EXERCISE 3.2
Lipids

Objectives

After completing this exercise, you should be able to

1. Define lipid and give examples.
2. Describe the test that indicates the presence of lipids.

Lipids are compounds that contain mostly carbon and hydrogen. They are grouped together solely on the basis of their insolubility in water. The lipids we will consider in this laboratory are fats and oils, which are generally used as storage molecules in both plants and animals. You are no doubt already familiar with the fact that your body converts excess food into fat. This fat is stored in your adipose tissue until your food intake is lower than your metabolic needs, at which time the fat can be metabolized to generate ATP, whose energy can be used for cellular work. Plants, too, can store fats. Seeds are often provisioned with fats that can be metabolized by the developing embryo when germination time comes. Thus we obtain corn oil, peanut oil, sunflower oil, and others by pressing the seeds.

You will use the **paper test** (Figure 3.4) to indicate the presence of lipids in various foods. Although this test is not very sophisticated, it is quick and convenient.

Figure 3.4.
Brown paper test for lipids.

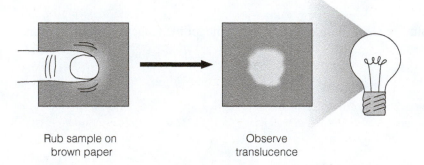

Rub sample on
brown paper

Observe
translucence

Procedure

1. Get a small square of brown paper. Write "oil" on one half and "water" on the other.
2. Put a tiny drop of salad oil on the half of the paper labeled oil. Rub it gently with your fingertip.
3. As a negative control, put a tiny drop of water on the half of the paper labeled water. Rub it gently with a different fingertip to avoid contamination.
4. Allow the spots to dry. This may take quite a while, so go on to another exercise while you wait.
5. When the spots are dry, hold the paper up to the light.

Interpretation of Results

Describe a positive test for lipids.

What are the limitations of this test?

E X E R C I S E   3 . 3
Proteins

Objectives

After completing this exercise, you should be able to

1. Define protein and give examples.
2. Explain why the structure of a protein is important for its function.
3. Describe the test that indicates the presence of protein.

A **protein's** structure is determined by the amino acid subunits that make up the molecule. Although there are only 20 different naturally occurring amino acids, each protein molecule has a unique sequence. The amino acids are linked by fairly tight bonds, and the side groups that are part of the amino acids also interact with each other to help shape the molecule.

Proteins have a greater diversity of roles than either carbohydrates or lipids. The shape of a protein is key to its purpose: Proteins work by selectively binding to other molecules.

You will use **biuret reagent** as a test for proteins (Figure 3.5). This reagent, which is blue, reacts with proteins to give a light violet or lavender color.

You will use a solution of egg albumin (a protein extracted from egg whites) to demonstrate a positive biuret test. What negative control should be used for this test?

Procedure

1. Get two test tubes and label them 1 and 2.
2. Put two droppersful of egg albumin into Tube 1.

Figure 3.5.
Biuret test for protein.

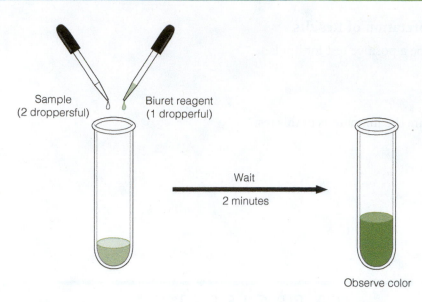

Sample
(2 droppersful)

Biuret reagent
(1 dropperful)

Wait
2 minutes

Observe color

3. Tube 2 is the control. What substance goes in it? How much should
    be used?

4. Put 1 dropperful of biuret reagent into each tube and swirl gently
    to mix.

5. After 2 minutes, record the color in each tube.
    Tube 1 (egg albumin):

    Tube 2 (negative control):

Interpretation of Results

Describe a positive biuret test.

What are the limitations of this test?

E X E R C I S E   3 . 4
Macromolecules in Food

Objectives

After completing this exercise, you should be able to

1. Explain how the components of butter are distributed in two fractions by the process of clarification.
2. Interpret the results of tests that indicate the presence of sugar, starch, lipid, and protein in an unknown sample.

We metabolize food in order to release energy to produce the ATP needed for cellular work. We also break down food molecules in order to use their subunits as raw materials for synthesizing our own macromolecules. In this exercise, you will investigate certain foods to learn which macromolecules are present in each.

Activity A: Separation of Butter

Most foods are complex mixtures of substances. Butter, for example, may appear to be solid fat, but it is actually a mixture of proteins, carbohydrates, and lipids. It is an emulsion, which means that the lipids occur in very small droplets dispersed throughout the water-soluble portion. (You will learn more about emulsions in Lab Topic 14, Digestion.)

The lipid can be separated from the water-soluble, protein-containing part of the butter in a process called clarification. Butter is often clarified for use in cooking. Once the water-soluble part has been removed, the lipid that remains can be used to fry at higher temperatures than for whole butter because it is the protein in butter that scorches first. It is also the protein that spoils most readily, so clarified butter keeps longer.

Procedure

1. Fill a 250-mL beaker approximately to the 75-mL mark with water.
2. Put the beaker on a hot plate and let the water come to a boil.
3. Cut approximately 1 tablespoon of butter into smaller chunks and put them in a 50-mL beaker.
4. When the water boils, remove the beaker from the hot plate with a potholder and set it on a paper towel on the lab bench. Choose a spot where it will not be disturbed.
5. Put the 50-mL beaker into the 250-mL beaker (water bath).
6. Leave the butter undisturbed for at least 15 minutes. While you wait, continue with steps 7–9.

7. Get approximately 1 teaspoon of butter from your instructor and put it into a test tube. (A teaspoon is $\frac{1}{3}$ of a tablespoon.)

8. Put the test tube in a warm water bath to melt the butter. (You can use the water bath the 50-mL beaker is in, but don't disturb the beaker.)

9. When the butter in the test tube has melted, perform the four tests that were introduced in this laboratory and record your results in Table 3.2. (The test procedures are reviewed in the next section.)

10. After 15 minutes, gently pick up the 50-mL beaker and set it in ice. Leave it undisturbed for about 10 minutes. Work on Activity B while the lipid (upper) layer solidifies.

11. The upper layer (clarified butter) should now be solid or semisolid. Remove it with a spatula and put it on a paper towel. Pat the bottom of the butter dry with a paper towel to remove any contaminants from the lower layer.

12. Place the clarified butter in a test tube and melt it in a warm water bath.

13. Perform the four tests on the melted clarified butter and record the results in Table 3.2.

14. Perform the four tests on the lower (liquid) layer of the clarified butter and record the results in Table 3.2. If you do not have enough liquid to perform all the tests, arrange to share results with another lab group.

Table 3.2

	Benedict's (sugar)	Iodine (starch)	Paper (lipid)	Biuret (protein)
Whole butter				
Clarified butter, upper layer				
Clarified butter, lower layer				

Interpretation of Results

Describe what happens to butter as a result of the clarification procedure.

Clarified butter lacks the "butter" flavor. What does this tell you about the molecules responsible for the taste of butter?

Some students try this procedure to clarify margarine. They warm the margarine and leave it on ice for 10 minutes as specified in the directions. They find that the margarine has resolidified but there is no liquid lower layer. What does this tell them about the margarine?

Activity B: Tests with Food

Test each of the items in Table 3.3 for the presence of simple sugars, starch, lipid, and protein. Your instructor may want to modify the list. The procedures for the tests are reviewed below.

Benedict's Test (sugar)

Put 1 pasteur pipetful of sample into a test tube. Add 2 droppersful of Benedict's reagent; mix. Heat in a boiling water bath for 5 minutes. Allow to cool and observe the precipitate.

 Some samples may require extra cooling time, so don't be too hasty in recording results.

Iodine Test (starch)

Put a pipetful of sample into a test tube and add 4 or 5 drops of iodine reagent; mix.

 In some foods, the starch is still contained in granules inside the cells. You may see these dark granules suspended in the yellow solution instead of seeing the entire solution turning blue.

Paper Test (lipid)

If the sample is whole (for example, a peanut), rub a piece of it directly on the paper. If the sample is liquid, put a small drop on the paper.

✳ Remember to wait for the paper to dry before you record the results.

Biuret Test (protein)

Put 1 pipetful of sample into the test tube and add 1 dropperful of biuret reagent; mix.

✳ Allow at least 2 minutes for the reaction to occur. Some samples may take 5 minutes to react.

Some of the foods to be tested are solids. Use a razor blade to mince approximately 1 cm^3 (about the size of a pea) of the sample. Put it in a test tube with 10 mL distilled water. Put your thumb over the top of the test tube and shake it vigorously for 1 minute. Perform the tests using the liquid (except the lipid test). Record your results in Table 3.3. Be sure to rinse off the razor blade and cutting board between samples to avoid contamination.

Table 3.3

	Benedict's (sugar)	Iodine (starch)	Paper (lipid)	Biuret (protein)
Banana				
Coconut				
Milk				
Peanut				
Potato				

Interpretation of Results

Which results confirmed your expectations about the composition of foods?

Which results were unexpected?

What factors might result in a false negative test (that is, the food does contain a molecule but the tests results are negative)?

Why might a plant storage organ (such as a fruit or tuber) contain both starch and sugar?

If you have tested foods in addition to the ones listed in Table 3.3, compare the results from those tests with the results for the foods listed in Table 3.3.

Questions for Review

1. What subunits make up
    a. Carbohydrates?

    b. Proteins?

2. Why is each test done initially using water as well as a known sample?

3. Why might a substance taste sweet, yet give a negative reaction with the Benedict's test?

What procedure could you use to check your answer to the previous question?

4. You have been given an unknown solution. Describe how you would test it for the presence of
   a. Starch:

   b. Lipid:

   c. Sugars:

   d. Protein:

5. You have tested an unknown sample with biuret and Benedict's reagents. The solution mixed with biuret reagent is blue. The solution boiled with Benedict's reagent is also blue. What does this tell you about the sample?

6. Whole butter gives only a slightly positive test for protein (and may show no reaction at all). When the same butter is clarified, however, the liquid lower layer is definitely positive for protein. Explain why these different results might have been obtained.

7. Since potatoes have starch in them, why don't they taste sweet after they are boiled?

Acknowledgments

Procedures for the macromolecule tests were adapted from the following sources:

Armstrong, W. D., and C. W. Carr. Physiological Chemistry Laboratory Directions, 3rd ed. Minneapolis: Burgess Publishing, 1963.

Dotti, L. B., and J. M. Orten. Laboratory Instructions in Biochemistry, 8th ed. St. Louis: C. V. Mosby, 1971.

Oser, B. L., ed. *Hawk's Physiological Chemistry*, 14th ed. New York: McGraw-Hill, 1965.

Using the Microscope

Introduction

The invention of the microscope was the first technological breakthrough in the study of biology. Although anatomists could investigate the basic architecture of organisms, very little could be learned about their composition before the microscope came into use. It made possible the discovery that all living organisms are composed of cells and that all cells are similar in some ways. It also enabled scientists to learn how various types of cells are different from each other. A light microscope can magnify specimens up to 1500×. Since a typical animal cell might be approximately 0.02 mm in diameter and the unaided eye can only distinguish objects as small as about 0.2 mm in diameter, the microscope is an essential tool for studying cells and subcellular components.

You will be using two types of microscopes, the compound light microscope and the stereoscopic dissecting microscope. Effective use of these microscopes is one of the fundamental skills required for studying biology, and it takes some practice to master this skill. This laboratory will show you how to use these instruments to their best advantage.

Outline

Exercise 4.1: The Compound Microscope

    Activity A: The Parts of the Compound Microscope

    Activity B: Using the Compound Microscope

    Activity C: Depth of Focus

Exercise 4.2: Observing Cells

    Activity A: Animal Cells

    Activity B: Plant Cells

Exercise 4.3: The Stereoscopic Dissecting Microscope

    Activity A: The Parts of the Dissecting Microscope

    Activity B: Using the Dissecting Microscope

Exercise 4.4: Observing an Animal Specimen

E X E R C I S E   4 . 1
The Compound Microscope

Objectives

After completing this exercise, you should be able to

1. Determine total magnification for the lenses on your microscope.
2. Explain the difference between magnification and resolving power.
3. Identify the parts of the compound microscope and give the function of each part.
4. Describe the procedure you would follow to locate and focus on any specimen using the scanning, low-power, or high-power objective.
5. Define field of view and describe how the field of view changes with magnification.
6. Define depth of focus and explain how it changes with magnification and how to adjust for it when viewing specimens.

The compound light microscope has two sets of lenses: the eyepiece, which is near the eye, and the objective, which is near the specimen. The total magnification of the specimen is calculated by multiplying the magnification of each set of lenses:

total magnification =
   magnification of objective × magnification of eyepiece

The eyepiece magnification is usually fixed, and ranges from 8 times (8×) to 15×, depending upon the microscope model. The objective magnification is changeable. Compound microscopes generally have three to five different objectives. Typical magnifications are 4× (scanning power), 10× (low power), 45× (high power), and 100× (oil immersion). The magnification is engraved on each objective.

Calculate the total magnification for the lens combinations on your microscope:

Scanning power: eyepiece _____× times objective _____× = _____×

Low power: eyepiece _____× times objective _____× = _____×

High power: eyepiece _____× times objective _____× = _____×

Oil immersion: eyepiece _____× times objective _____× = _____×

Although theoretically we could put even higher powered lenses on microscopes to achieve greater magnifications, there is a practical limit imposed by the resolving power of the optical system. The **resolving power** of a microscope is its ability to distinguish two objects that are very close together as separate; it is what allows us to see detail.

When two objects cannot be distinguished as separate, the image looks blurry. Resolving power is thus a measure of clarity. Two microscopes may give the same magnification, yet the clarity of images provided by one may be better due to superior resolving power, which in turn is a function of the quality of the objective lens.

Activity A: The Parts of the Compound Microscope

Although there are numerous variations in the features of different models of compound microscopes, they are all constructed on a common plan. Figure 4.1 shows a typical model.

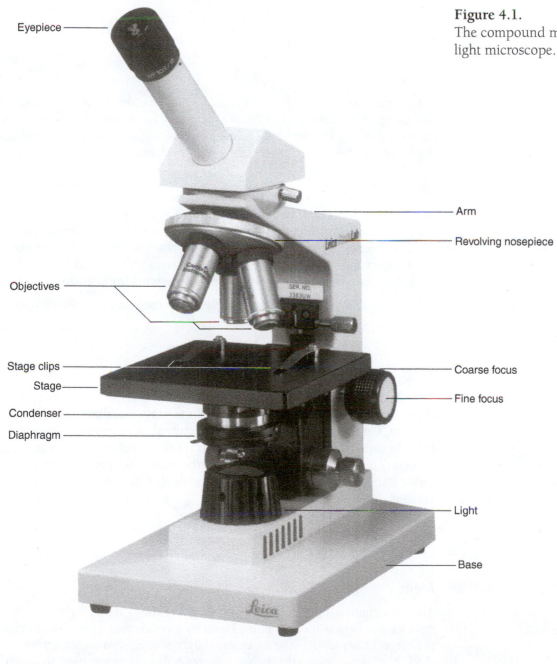

Figure 4.1.
The compound monocular light microscope.

Eyepiece

Arm

Revolving nosepiece

Objectives

Stage clips

Stage

Condenser

Diaphragm

Coarse focus

Fine focus

Light

Base

Locate the following parts on your microscope. The descriptions of parts include the common variations found in different models of microscopes. In the space below each item, answer the questions about the microscope you will be using.

Eyepiece

The eyepiece, also called the ocular, is the topmost lens system, and serves to magnify the image of the specimen. Some models have one eyepiece (monocular). When using a monocular scope, it is recommended that you leave the other eye open (if you can master this technique) or else cover the other eye with one hand, rather than squinting with one eye tightly shut. Binocular (two-eyepiece) models are also available.

How many eyepieces does your microscope have?

Is it monocular or binocular?

On a binocular microscope, the eyepieces can be moved closer together or farther apart to match the distance between the user's pupils. There is sometimes also a diopter ring on one eyepiece, which allows you to adjust the focus for that eye before focusing with both eyes.

Objectives

The objectives magnify and resolve the image of the specimen. They are set in a revolving nosepiece so that the user can readily change lenses. Most microscopes have a ring above the lenses. Use this to turn the nosepiece rather than grabbing the lenses themselves. You should feel the lens click into place when it is engaged. This is the **working position.**

What are the magnifications of the objective lenses on your microscope?

Stage

The slide holding the specimen is placed on the stage and secured with clips. There is a hole in the center of the stage to allow light to pass through. On most microscopes, you move the slide around the stage manually in order to locate areas of interest on the specimen. Other models, however, have a movable stage. Once the slide is clipped in place, you rotate knobs to move the slide from side to side or toward or away from yourself.

Does your microscope have a movable stage?

Light

The specimen is illuminated by substage lighting. If the light source is built in, the light is focused into a beam by a **condenser** located just below the stage. Other models use an external light source; you adjust a mirror to direct the light upward through the stage. A **diaphragm** is used to control the aperture (opening), or width of the beam of light that reaches the specimen. Some microscopes allow the user to vary the aperture continually by using a lever. Other models have a wheel with fixed apertures of various sizes; you turn the wheel to select the appropriate size.

Describe the illumination system for your microscope:

Focus Knobs

Once the slide containing the specimen is mounted on the stage, you look through the eyepiece and adjust the focus using first the coarse-focus knob (to bring the specimen into view) and then the fine-focus knob (to sharpen the image). On most microscopes, turning a focus knob toward you raises the objective, and turning it away from you lowers the objective. On some microscopes, however, the stage, rather than the objectives, moves up and down.

Describe how the focus knobs on your microscope work:

Support Structure

The body of the microscope is constructed to protect the lenses and the focusing system and to provide a sturdy, vibration-free support stand.

❋ **You should never attempt to remove the lenses or any other part of the microscope.**

When working with a microscope, observe the following rules:

1. Carry the microscope with two hands. Hold the arm with one hand and use your other hand to support the base.
2. Keep all parts of the microscope dry. If you spill anything on the stage, wipe it up immediately.
3. If the lenses need to be cleaned, always use lens paper. Never use paper towels, tissues, or cleaning tissues. They will scratch the lenses.
4. When you are finished using the microscope, remove the slide and turn the lowest power objective to the working position.

Activity B: Using the Compound Microscope

Before you begin using the microscope to examine unfamiliar creatures, you will first do an exercise using a recognizable object, the letter "e," to help you learn to use the microscope effectively. The keys to successful microscope work are locating the specimen, focusing on the specimen, and adjusting the light for the best contrast and resolution.

* If you wear glasses, you should remove them to look through the microscope unless they correct an astigmatism.

Procedure

1. Turn on the light source and make sure the lowest power (scanning power, if available) objective is in the working position.

2. Place the "e" slide on the stage, centering the "e" over the hole in the stage. Secure the slide with the clips.

3. Look at the slide on the stage from the side (*not* through the eyepiece) and turn the coarse-focus knob until the objective is as close to the stage as it will go. (Most microscopes have a built-in stop to prevent you from crashing the lens into the stage, but be careful in case yours doesn't.)

4. If your microscope is binocular, adjust the distance between the eyepieces by pushing them closer together or moving them farther apart until you can comfortably see one image. If there are markings on the microscope, record the setting so you can preset it in the future:

5. Look through the eyepiece(s) and slowly move the stage and objective farther apart by turning the coarse-focus knob until you have the "e" or some part of it in focus. Move the slide to center the "e" in your **field of view** (the area of the image that you see).

* You may see a black line in the field of view. This is a pointer. It allows you to indicate a structure that you want someone else to view.

6. If you are using a monocular microscope or a binocular microscope that doesn't have a focusing ring for one eyepiece, use the fine-focus knob to bring the image into sharp focus.

   If you are using a binocular scope that has a focusing ring, first look only through the eyepiece that doesn't have the focusing ring. If the focusing ring is on the left eyepiece, look through the right eyepiece and cover the left eyepiece. Focus the "e" for that eye using the fine-focus knob. Then look through the eyepiece with the focusing ring, keeping the other eye covered. Adjust the focusing ring until the image is sharp. Finally, view the image with both eyes.

   How is the image of the "e" oriented compared to the image of the "e" on the slide itself?

   Look through the eyepiece and move the slide to your right.

What happens to the "e"?

What happens when you move the slide toward you?

7. Open the diaphragm all the way (or set the disk to the largest aperture), and then slowly close it (or choose smaller and smaller apertures in the disk). You can see that the amount of light affects the image you receive. In general, you should use a smaller opening for scanning and low-power objectives and a larger opening for high-power and oil-immersion objectives. However, the type of specimen also affects how much light is needed, so it is best for you to try different settings. Begin with the largest aperture and reduce the aperture until the image has good contrast, but you are still able to see details.

8. Bring the low-power objective into the working position. If your microscope is parfocal, you should only need to adjust the focus a little using the fine-focus knob. **Parfocal** means that when an image is focused with one objective, it will also be in focus in the other objectives. Center the "e" in your field of view.

   Can you still see the entire "e"?

9. Now turn the high-power objective to the working position, being careful not to bang the objective on the stage clips as you turn it.

> ✳ This objective leaves very little space between the slide and the objective. You should use only the fine-focus knob when you are on high power.

Describe what you see at this magnification.

How does the field of view at low power compare to the field of view at high power?

Steps 10–15 will give you an idea of the relative fields of view that you see with your microscope.

10. Put the scanning (or lowest power) objective in the working position.

11. Turn the coarse-focus knob so that it is as close to the stage as it will go.

12. Place a transparent ruler across the stage and focus on the edge that has millimeter markings.

✳ **If you can't bring the markings into focus, put a clean glass slide under the ruler to raise it closer to the lens.**

---

13. Move the ruler so that one of the markings is at the very edge of your field of view (see Figure 4.2). Each marked interval on the ruler is 1 mm. Estimate the diameter of the field of view at this magnification.

Figure 4.2.
Alignment of ruler to measure the field of view.

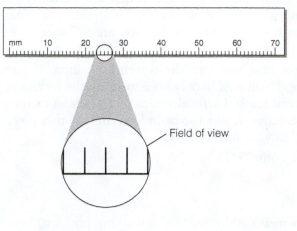

Magnification: _____

Diameter of field of view: _____

14. Change to the next higher objective and estimate the diameter of the field of view.

Magnification: _____

Diameter of field of view: _____

✳ **If your microscope only has low-power and high-power objectives, omit step 15.**

---

15. Measure the field of view again at the next higher objective.

Magnification: _____

Diameter of field of view: _____

Obviously, microscopic images are too small to be measured with a ruler. Instead of expressing measurements in millimeters, they are usually expressed as micrometers (μm). $1 \ \mu m = 1 \times 10^{-6}$ m. Since $1 \ mm = 1 \times 10^{-3}$ m, $1 \ mm = 1 \times 10^{3} \ \mu m$. For example, to convert 5 mm to μm:

$$5 \ mm \ (1 \times 10^{3} \ \mu m/mm) = 5 \times 10^{3} \ \mu m = 5000 \ \mu m$$

If you are not familiar with these units or working with scientific notation, see Appendix A.

Convert the field of view measurements you recorded above to micrometers:

Magnification: _____

Diameter of field of view: _____

Magnification: _____

Diameter of field of view: _____

Magnification: _____

Diameter of field of view: _____

Why is this technique for measuring the field of view not useful for an oil immersion lens?

Activity C: Depth of Focus

Although a specimen mounted on a slide appears to be completely two-dimensional, there is actually 3 to 10 μm of depth. That might not sound like a lot, but keep in mind that the specimens are magnified up to 1000×. At higher magnifications, the lens will not focus on the entire specimen at once (Figure 4.3). It is therefore useful to focus up and down through the entire depth of a specimen using the fine-focus knob, especially when you are using high power. The thickness of the specimen that can be seen in focus at any time is called the **depth of focus.** You will investigate depth of focus in the following procedure.

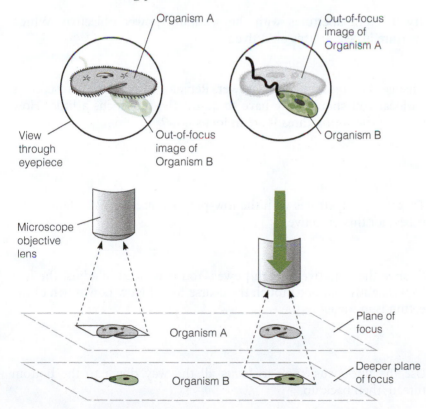

Figure 4.3.
Depth of focus. When a specimen is thick, only part of it can be in focus at a time.

You will also learn the technique for making a **wet mount,** which simply means that you place your specimen in a drop of water (or another liquid) on a slide and add a coverslip. The coverslip flattens the specimen, giving the slide a uniform depth for easier viewing, and protects the lens from getting wet.

Procedure

1. Get a clean microscope slide. With forceps, pull a *tiny* amount of cotton off of a cotton ball and put it on the middle of the slide.

2. Put a drop or two of *Paramecium* culture onto the cotton. (Hint: get a drop from the bottom of the container.)

3. Add a coverslip, using a dissecting needle (as shown in Figure 4.4) to lower the coverslip onto the slide. The trick is to do it slowly enough to keep any air bubbles from being trapped under the coverslip.

Figure 4.4.
Making a wet mount.

4. Put the slide on the stage of the microscope and focus using scanning power. Of the cotton threads that appear in your field of view, how many are in focus?

5. Try different apertures with the scanning power objective. Which aperture is best for this objective?

6. Change the objective to low power. Remember, if your microscope is parfocal you should only have to adjust the fine focus a little. How many of the cotton threads are in focus now?

    Try different apertures with the low-power objective. Which aperture is best for this objective?

7. Change the objective to high power. You may need to adjust the fine focus slightly, but don't touch the coarse focus! Now how much of the cotton is in focus?

    Use the fine-focus knob to focus all the way down to the bottom thread, then back up.

8. Try different apertures with the high-power objective. Which aperture is best for this objective?

9. Locate a *Paramecium* and watch it swim among the threads.

E X E R C I S E   4 . 2
Observing Cells

Objectives

After completing this exercise, you should be able to

1. Explain what stains are and why they are often used with biological materials.
2. Draw a cheek cell and identify its nucleus and plasma membrane.
3. Draw an *Elodea* cell and identify the chloroplasts and cell wall.

Activity A: Animal Cells

Animal cells contain numerous organelles—such as Golgi bodies, endoplasmic reticulum, lysosomes, microbodies, and mitochondria—that are too small to be seen with the compound light microscope. The only animal organelle that is reliably visible by compound light microscopy is the nucleus, which may be 4–6 µm in diameter. You can also tell where the plasma membrane is (though you can't see it, since it's less than 0.01 µm thick) because it forms a boundary around each cell.

Most animal cells are relatively colorless and appear transparent under the light microscope. We can tag certain structures by attaching pigment molecules to them in a procedure called **staining.** In this exercise you will use the stain methylene blue, which dyes the acidic molecules of the nucleus (the nucleic acids) blue. It also imparts a light blue tint to the rest of the cell.

Procedure

1. Gently scrape some cheek cells from the inside of your cheek with a clean, flat toothpick.

 Discard your toothpick immediately after use in the container provided.

---

2. Spread the scrapings in the middle of a clean slide. Wait until the slide dries before proceeding.

3.   Set half of a small petri dish inside half of a large petri dish. Place the slide on top of the small petri dish (see Figure 4.5).

Figure 4.5.
Arrangement of petri dishes for staining cheek cells. The small petri dish serves as a platform for the slide, and the large petri dish will catch the excess water that washes off the slide.

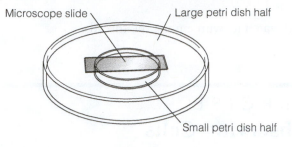

Microscope slide   Large petri dish half

Small petri dish half

4.   Put several drops of methylene blue on the cheek scrapings. Wait 2 minutes for the stain to take effect.

5.   *Gently* rinse the methylene blue off the slide with water from a squirt bottle. (If you rinse too vigorously, you may wash the cells off, too.)

6.   Use a paper towel to blot dry the bottom of the slide. Do not wipe the top!

7.   Add a small drop of water to the cheek scrapings; then add a cover-slip.

8.   Examine your cheek cells under the compound microscope.

> **Since the cells are still relatively transparent, adjust the aperture of the diaphragm to obtain maximum contrast.**

9.   Sketch the cheek cells in the margin of your lab manual and label the plasma membrane and nucleus.

If your microscope is equipped with an oil immersion lens, complete steps 10–14.

10.   Be sure your slide is focused and centered at high power.

11.   Move the high-power objective out of the way so that no objective is directly over the slide.

12.   Put a small drop of immersion oil on the coverslip in the area you wish to view.

13.   Turn the ring above the objectives so that the oil immersion lens is in the working position. When it is in place, the lens will actually be in the oil drop—this is why it's called oil immersion. View the specimen. You will need to open the diaphragm as wide as possible to get enough light. Sketch a cell in the margin of your lab manual.

What has happened to the size of your field of view?

14.   When you are finished, wipe the oil immersion lens with lens paper to remove the oil.

Your cheek cells are a type of tissue called **squamous epithelium,** the same kind of tissue that makes up your skin and other internal and external linings of your body. The cells are loosely joined together in sheets (see Figure 4.6).

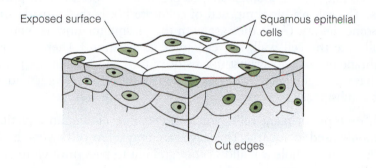

Exposed surface   Squamous epithelial cells

Cut edges

Figure 4.6.
Squamous epithelium.

The specimen you prepared showed cells lying flat on the slide, and you looked down on top of them through the microscope. Most tissues, however, don't peel off in sheets the way squamous epithelium does. For example, if you want to examine the skin tissues that lie beneath the squamous epithelium, you have to make a very thin slice from a block of tissue so that light can pass through it (see Figure 4.7). The tissue is first fixed, or killed, and then sliced and stained. When the specimen is mounted on a slide, a coverslip is fixed in place permanently. This type of slide is called a prepared slide, in contrast to the wet mounts you make yourself. You will frequently use prepared slides in this course.

Procedure

View a prepared slide of skin. How can you identify the squamous epithelium?

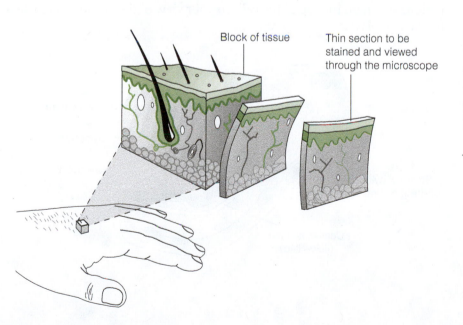

Block of tissue   Thin section to be stained and viewed through the microscope

Figure 4.7.
Preparation of a thin section of tissue. After a block of tissue is removed from a larger specimen, it is sliced thinly enough to allow light to pass through it.

Activity B: Plant Cells

In addition to the nucleus, chloroplasts and the cell wall are visible when you look at a plant cell under the light microscope. Chloroplasts are the large (~8 μm), green, football-shaped organelles responsible for photosynthesis. The cell wall is composed of cellulose fibrils that provide strength and some rigidity. (As is the case with the plasma membrane, you cannot actually see the cell wall, but can infer its position.) There is a plasma membrane pressed up against the inside of the cell wall. Mature plant cells also have a large central vacuole, which is filled with water and water-soluble substances and may occupy 90% of the cell volume.

As an example of plant cells, you will examine *Elodea*, an aquatic plant commonly used in aquaria. The leaves are only a few cell layers thick, so a wet mount of a whole leaf allows enough light to pass through to examine the cells with light microscopy.

Procedure

1. Place one *Elodea* leaf on a slide in a drop of water.
2. Add a coverslip as shown in Figure 4.4. Try to avoid trapping air bubbles on the leaf surface.
3. Focus on the *Elodea* cells at scanning power, then at low power, and finally at high power.
4. With the high-power objective in place, move the fine-focus knob slightly up and down to see the entire thickness of a cell.

Sketch a few *Elodea* cells in the margin of your lab manual. Label the cell wall and chloroplasts. (You may not be able to see the nucleus.) In some cells you should be able to see cytoplasmic streaming as the chloroplasts move around the cell on microfilament tracks.

Try to visualize the *Elodea* cell in three dimensions. The cell is shaped like a shoebox, and inside the cell wall is the plasma membrane. A thin layer of cytoplasm lines the cell, but the largest volume is occupied by the central vacuole, which is interior to the cytoplasm. As Figure 4.8 shows, you are looking down on the top of the cell, so your view of the vacuole is blocked by the cytoplasm.

Figure 4.8.
Three-dimensional view of *Elodea* cell.

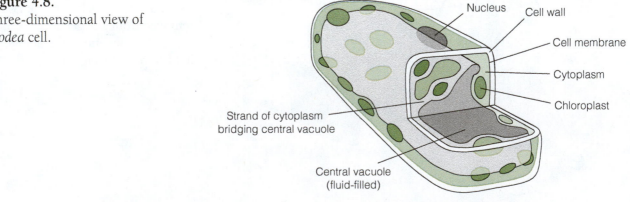

Nucleus
Cell wall
Cell membrane
Cytoplasm
Chloroplast
Strand of cytoplasm bridging central vacuole
Central vacuole (fluid-filled)

E X E R C I S E   4 . 3
The Stereoscopic Dissecting Microscope

Objectives

After completing this exercise, you should be able to

1. Explain how a dissecting scope differs from the compound microscope.
2. Identify the parts of the dissecting scope and give the function of each part.
3. Describe how to locate and focus on a specimen using the dissecting scope.
4. Identify situations in which a dissecting scope would be more useful than a compound light microscope.

In order to be viewed with the compound light microscope, specimens must be thin enough for light to pass through them. They must be mounted on glass slides; usually they are covered by a coverslip, so little manipulation of the specimen is possible while it is being viewed. In addition, the minimum magnification is typically 40×.

The stereoscopic dissecting microscope, usually referred to as a dissecting scope, provides a lower range of magnifications, usually between 5× and 50×, and a large working distance for manipulation of specimens. This type of microscope is also a compound microscope, since there are two lens systems, but in this case the objective lenses are housed inside the scope and are changed using a knob or dial. The number and magnifications of objective lenses vary considerably from one model of microscope to another. Some have only two settings, low power and high power, while others can be varied continuously like a zoom lens.

All dissecting scopes have two eyepieces. In contrast to a binocular compound microscope, in which you see the same image with both eyes, in the dissecting scope each eye sees through an independent system—there are actually two objectives. The result is that you see depth the way you do with unaided eyes.

Activity A: The Parts of the Dissecting Microscope

Locate the following parts on your dissecting scope (see Figure 4.9).

Eyepieces (oculars)

These lenses magnify the specimen. Notice that the distance between the eyepieces can be adjusted to fit the distance between your eyes. Often one eyepiece has a diopter ring on it, which allows you to adjust the focus of the two eyepieces independently.

What is the eyepiece magnification on your dissecting scope?

Stage

Usually the specimen may be placed directly on the stage to be examined. A slide is required only for some types of specimens (for instance, if the specimen must be viewed in a drop of water). A hole in the center of the stage allows light to be transmitted from beneath the stage.

Magnification Changer

This dial or knob changes the objective lenses inside the microscope. It is marked to indicate the magnifications possible for the objectives. There may be a setting that is less than 1 (for example, 0.7). If the eyepiece is 10×, this objective would give a total magnification of 7×.

What are the objective magnifications on your dissecting scope?

What are the total magnifications for your dissecting scope?

Focus Knob

Note that there is only one type of focus knob on this kind of microscope.

Light Sources

Two types of lighting are used with dissecting scopes: transmitted light, which comes from beneath the stage, and reflected light, which shines down on the specimen from above. Transmitted light is used when the specimen is thin and transparent. Reflected light is used for specimens that are opaque. It enhances the three-dimensional quality of the image.

Describe the illumination system in your microscope.

Activity B: Using the Dissecting Microscope

Familiarize yourself with the operation of the dissecting scope by using the "e" slide again.

Procedure

1. Position the "e" slide on the stage so that the "e" is directly over the hole in the center of the stage.
2. Set the objective magnification at its lowest power.
3. Turn on the transmitted (substage) light.
4. Focus on the "e." (You may need to complete steps 5 and 6 to get the best image.)
5. While looking through the eyepieces, move them together or apart until you see a single image. If there are markings on the microscope, record the setting for your intereye distance for future reference:

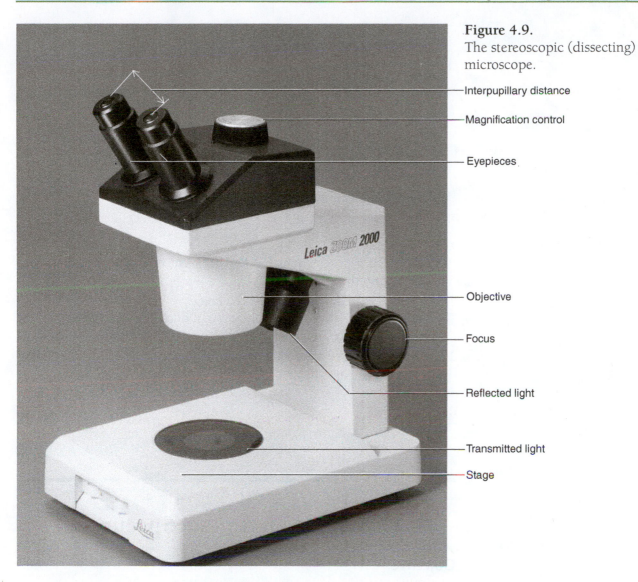

Figure 4.9.
The stereoscopic (dissecting) microscope.

— Interpupillary distance

— Magnification control

— Eyepieces

— Objective

— Focus

— Reflected light

— Transmitted light

— Stage

6. With one eye, look through the eyepiece that doesn't have the focusing (diopter) ring. Cover the other eye. Use the focus knob to bring the image into sharp focus. Then look through the eyepiece that has the focusing ring and cover the other eye. Turn the ring until you get the sharpest focus for that eye. Then look with both eyes. The image should be in focus.

    How is the image of the "e" oriented compared to the image of the "e" on the slide itself?

    Look through the eyepiece and move the slide to your right.
    What happens to the "e"?

    What happens when you move the slide toward you?

7. Turn the magnification knob so that you see the "e" through the entire range of magnifications available.

8. Turn off the transmitted light and turn on the reflected light. How does this change the image you see?

Which light source is better for this slide?

9. Next, get a photograph that has been cut out of a newspaper and place it on the stage.

10. Using reflected light, focus on the image. When you magnify the picture, you can see that it is composed of dots. The individual dots are not resolved by the unaided eye.

11. Turn off the reflected light and turn on the transmitted light. How does this change the image you see?

E X E R C I S E   4 . 4
Observing an Animal Specimen

Objective

After completing this exercise, you should be able to

1. Determine the best illumination and magnification to use for viewing a specimen on the dissecting scope.

As you have seen, the dissecting scope is the best choice if you are working with large specimens requiring relatively little magnification. In this exercise, you will examine a hydra, an animal related to sea anemones and corals. The hydra is commonly found in ponds and streams and is a favorite laboratory organism for introductory biology.

Procedure

1. Get a prepared slide of a hydra and place it on the stage of the dissecting scope.

2. Use a magnification that allows you to see the entire hydra and focus on it.

    Which type of light gives you the best image?

3. Sketch the hydra in the margin of your lab manual.

4. Get a dish containing living hydra. Place it on the stage and focus on one hydra.

   Which type of light gives you the best image?

   Describe how the image of the living hydra is different from the image of the prepared slide.

5. Dip a piece of thread in the juice from liver. While you are looking at the hydra under the dissecting scope, place the thread in the dish near the hydra. Describe the animal's behavior.

Your instructor will provide other specimens for viewing under the dissecting scope.

Questions for Review

1. If the eyepiece on a microscope has a magnification of 10×, what is the total magnification with a 10× objective?

   What is the total magnification with a 45× objective?

2. A microscope gives a total magnification of 1500×, but the image is too blurry to be useful. What might be the problem with this microscope?

3. Why is it important to center a specimen on low power before attempting to focus on it at high power?

4. A student focuses on a specimen at low power and carefully centers it before changing to high power. At high power, however, he doesn't see the part of the specimen he was interested in. What might be the problem?

5. Inspired by her biology lab, a student decides to make a closer study of the food she eats. She uses a razor blade to make a very thin section from a raw potato and mounts it in a drop of water on a slide. To her disappointment, she can barely make out the cells under the microscope. What might she do to improve her results?

6. Compare the image obtained using a compound light microscope with the image obtained using a stereoscopic dissecting microscope.

Enzymes

Introduction

Biological processes depend on molecular catalysts called **enzymes** to speed up the chemical reactions that are necessary for cells to function. It takes energy to initiate some chemical reactions. Enzymes work by lowering that amount of energy so it's easier for the reaction to get started. Without enzymes the reactions would take place far too slowly to support life. The basic components of an enzyme-catalyzed reaction are the substrate(s), the product(s), and the enzyme itself.

The **substrate** is the reactant molecule that is changed by the enzyme. Each enzyme has a specific substrate or type of substrate. For example, the substrate sucrose is acted upon by the enzyme sucrase. In the reaction shown in Figure 5.1, a disaccharide, sucrose, is broken down into its component monosaccharides glucose and fructose, which are the **products** of this reaction. The enzyme itself is neither changed nor destroyed during the reaction.

Enzymes speed up the synthesis of biological molecules as well as the breakdown, but synthesis is usually more complex than breakdown and requires a series of reactions. When sucrose is synthesized from glucose and fructose, for example, there are six steps to the process. Each step is catalyzed by a different enzyme.

There are many chemical reactions that are common to most organisms, so the enzymes that catalyze these reactions are also common. For example, the process by which energy is harvested from glucose molecules is nearly universal, so almost all organisms have common enzymes that are used in this common pathway However, the specific properties and behaviors of enzymes may be different in different organisms. Even within the same organism, the version of an enzyme found in one organ may be slightly different from the version found in a different organ.

In addition to differences in the enzyme molecules themselves, the environment in which an enzyme works is an important influence on the reaction rate.

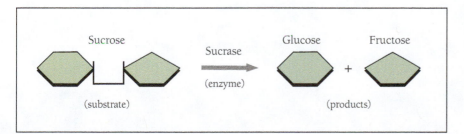

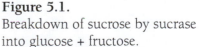

Figure 5.1.
Breakdown of sucrose by sucrase into glucose + fructose.

In this lab topic you will learn a method for investigating catecholase, an enzyme that is found in many plants. Your lab team will then design and perform your own experiment using this method.

Outline

Exercise 5.1: Factors Affecting Reaction Rate

Activity A: Quantity of the Reactants

Activity B: Physical Factors

Exercise 5.2: The Catecholase-Catalyzed Reaction

Exercise 5.3: Designing an Experiment

Exercise 5.4: Performing the Experiment and Interpreting the Results

E X E R C I S E   5 . 1
Factors Affecting Reaction Rate

Objectives

After completing this exercise, you should be able to

1. Make predictions regarding the effects that the amounts of enzyme and substrate present will have on the rate of an enzyme-catalyzed reaction.

2. Make predictions regarding the effects that physical factors such as pH, salt concentration, and temperature will have on the rate of an enzyme-catalyzed reaction.

Activity A: Quantity of the Reactants

Enzyme molecules do not undergo permanent changes during a reaction. After "turning over" one substrate molecule into product, the enzyme is free to engage with another substrate molecule and repeat the process. Enzymes work at a steady pace, turning over substrate into product for as long as substrate is available.

If substrate is abundant, what should happen to the reaction rate (amount of product formed/unit time) when **more enzyme** molecules are added to the reaction mixture? Sketch your prediction on the axes in Figure 5.2. Remember to graph the independent variable on the horizontal axis and the dependent variable on the vertical axis. Label both axes.

Figure 5.2.
Effect of enzyme concentration on reaction rate.

If the number of enzyme molecules is constant and the number of substrate molecules is low, what do you expect to happen to the reaction rate when **more substrate** is added to the reaction mixture? Sketch your prediction on the axes in Figure 5.3 and label both axes.

Figure 5.3.
Effect of substrate concentration on reaction rate.

Activity B: Physical Factors

Most enzymes are protein molecules. Recall that proteins are composed of a string of amino acids linked together with relatively tight bonds. The amino acids have different chemical groups attached to them. Some of these groups attract or repel each other, and some are hydrophobic or hydrophilic. As a result of these interactions, the protein molecule is folded into a three-dimensional shape, which is closely related to the protein's function. In the case of enzymes, the shape of the protein molecule is what enables it to establish the close association with the substrate molecule(s) that is the prerequisite for making the reaction happen. The particular part of the enzyme molecule where the substrate fits in is called the **active site.** Thus anything that alters the shape of the enzyme molecule, especially at the active site, may prevent it from functioning correctly.

The attractions and repulsions between chemical groups on the amino acid chain are strongly affected by **pH,** which is a measure of the concentration of hydrogen ions in solution (Lab Topic 2, pH and Buffers). Each enzyme has a pH at which it is most perfectly shaped; it therefore functions best at that pH. If the pH is too low (acidic) or too high (basic), the enzyme molecule loses its shape and thus its ability to catalyze reactions.

How do you think reaction rate will change in solutions of different pH? Sketch your idea on the axes shown in Figure 5.4 and label both axes.

Figure 5.4.
Effect of pH on reaction rate.

Salt in the enzyme's environment can also cause distortion of its shape by changing the interactions of the chemical groups on the amino acids. Just as with pH, there is an optimum salt concentration at which the enzyme is shaped perfectly to engage the substrate.

Temperature also has a profound effect on enzyme reactions. Since the enzyme molecules must be in actual contact with substrate molecules for the enzyme to catalyze the reaction, anything that increases the number of collisions between enzyme and substrate is expected to increase the reaction rate. Sketch this relationship on the axes in Figure 5.5 and label both axes.

Figure 5.5.
Effect of temperature on reaction rate.

But heat energy has another effect in addition to speeding up the movement of molecules. It can also disrupt the delicate attractions between the chemical side groups of amino acids, causing the enzyme to change its shape and perhaps even be denatured. So when the temperature is too high, the enzyme can't catalyze the reaction. Therefore, like pH and salt concentration, enzymes work best in an optimum temperature range. Adding this information to the graph you just drew, sketch how temperature is related to reaction rate on the axes in Figure 5.6.

Figure 5.6.
Effect of temperature on reaction rate.

E X E R C I S E 5 . 2
The Catecholase-Catalyzed Reaction

Objectives

After completing this exercise, you should be able to

1. Name the substrate, enzyme, and product in the experiment in Exercise 5.2.
2. Explain how the catecholase reaction is measured in the experiment in Exercise 5.2.

In order to study how various factors affect a particular enzyme, we need to be able to measure either the disappearance of substrate or the appearance of product. In this laboratory you will study the enzyme **catecholase**, which catalyzes a reaction in which **catechol**, the substrate, becomes the product **benzoquinone**. Benzoquinone is a reddish-brown color, so we can easily determine how much benzoquinone has been formed. In fact, you are already familiar with this reaction: You are observing it when you see the cut surface of an apple or potato turn brown. A good source of catecholase is an extract made from potatoes.

Figure 5.7.
Catecholase reaction. The colorless substrate is converted to a reddish-brown product.

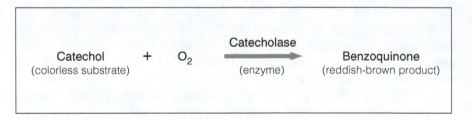

Catechol + O$_2$ →[Catecholase] Benzoquinone
(colorless substrate)   (enzyme)   (reddish-brown product)

Since we can use color to visualize product formation, we need a means of measuring how much color change happens during the reaction. One way to do this is to use an instrument called a spectrophotometer. A common model, and the one referred to in this lab, is the Spectronic 20 (Spec 20—see Figure 5.8). You may have a different model in your laboratory.

Figure 5.8.
Spec 20.

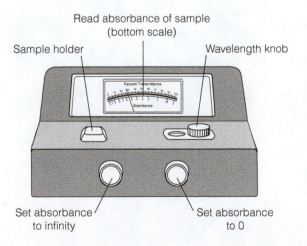

Read absorbance of sample (bottom scale)

Sample holder

Wavelength knob

Set absorbance to infinity

Set absorbance to 0

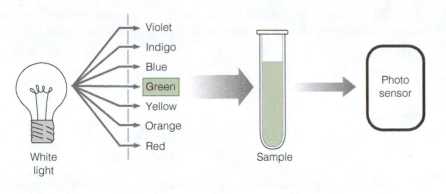

Figure 5.9.
Diagram of how a Spec 20 works. Like a prism, a spectrophotometer divides white light into its component wavelengths. In this diagram, green light has been selected to shine through the sample.

The Spec 20 measures catecholase activity by measuring the color change in reaction mixtures. As shown in Figure 5.9, this instrument shines light through a sample of reactants in a test tube and measures the amount of light that penetrates through the tube. This tells us how much light was absorbed by the sample, which in turn is a measure of how much product has been formed. The more of the reddish-brown product that has been made, the more light will be absorbed.

If spectrophotometers are not available, you may use a color chart to determine how much reaction has occurred. Compare your samples with Plate 1 and record the number for color intensity that matches the sample most closely. The more intense the color, the more benzoquinone has been formed.

On the axes in Figure 5.10, sketch how absorbance or color intensity (y-axis) varies in relation to product formation (x-axis) for the catecholase reaction. Label both axes.

Figure 5.10.
Relationship of product formation and absorbance.

In order to make the reaction occur, the mixture must contain both substrate and enzyme. Name the specific components for this reaction.

Substrate:

Enzyme:

You will set up three sample tubes to illustrate how this method works. If you are using a Spec 20 or other spectrophotometer, you must first zero the instrument using a blank. (If you are using the color chart method, you may skip the rest of this paragraph.) The blank is a type of control. You will notice that the potato extract itself has color; it already absorbs some light. That amount of light must be subtracted from the amount of light absorbed by the product of the reaction. This procedure is analogous to taring a balance. The blank contains enzyme (potato extract) but no substrate, so very little reaction will take place. (There will be some reaction, though, because the potato extract itself contains substrate.)

Look at Table 5.1 to see how the experiment has been designed and answer the following questions.

Table 5.1
Reaction Mixtures for Catecholase Experiment

Tube	Water (mL)	Catechol (mL)	Extract (mL)
1	3	2	1
2	3.5	2	0.5
3	4	2	0

What hypothesis is being tested?

What is the independent variable in this experiment?

What is the dependent variable?

Is there a control for this experiment?

Predict the outcome of your experiment in terms of your hypothesis. Explain what results would support your hypothesis and what results would prove your hypothesis false.

Procedure

Zeroing the Spec 20 (skip steps 1–9 if you are using the color chart)

1. Get a clean test tube (either a small test tube or a special Spec tube) and use a wax pencil to label it B (for blank). If the tube does not already have a short vertical mark at the top, draw one with wax pencil. When you put the tube into the sample holder, the mark should always face front.
2. Measure 1 mL potato extract into Tube B.
3. Add 5 mL distilled water to Tube B.
4. Cover Tube B tightly with Parafilm and invert it to mix the contents.
5. Set the wavelength knob on the Spec 20 to 540 nm (see Figure 5.8).
6. Using the knob on the left (Figure 5.8), set the absorbance reading (bottom scale) to infinity.
7. Wipe Tube B with a cleaning tissue and insert it into the sample holder of the Spec 20 with the vertical mark facing front.
8. Using the knob on the right, set the absorbance reading to 0.
9. Set Tube B aside. You will need to rezero the Spec 20 later.

The Spec 20 is now zeroed and ready to measure the tubes in which the catecholase reaction has occurred.

Measuring Color Change in Samples

10. Use a pipet to measure the water into Tubes 1, 2, and 3 (see Table 5.1).
11. Use a different pipet to add 2 mL catechol to each tube.
12. Use a third pipet to add 1 mL potato extract to Tube 1. Add 0.5 mL potato extract to Tube 2. Do not add any potato extract to Tube 3.
13. Place Parafilm tightly over the top of each tube and invert the tubes to mix their contents.
14. Shake each tube gently at 1 minute intervals to keep the reactants well mixed.

Spec 20 Method

15. After 3 minutes, use the blank to rezero the Spec 20 (follow steps 6–8).
16. Insert each tube in turn into the sample holder and read the absorbance (bottom scale). Record the absorbance reading for each tube in Table 5.2.

Color Chart Method

15. After 5 minutes, compare each tube to the catecholase chart on Color Plate 1. Observe the intensity of the colors (the intensity is more important than the actual color). In Table 5.2 record the number of the color intensity that most closely matches the color intensity of each tube.

Table 5.2
Results of Catecholase Experiment (absorbance or color intensity is recorded after 3 minutes)

	Tube		
	1	2	3
Absorbance/color intensity			

Was your hypothesis proven false or supported by the results? Use data to support your answer.

Predict the color change of a tube containing 3.75 mL water, 2 mL catechol, and 0.25 mL of potato extract. Explain how you derived this prediction from your data.

Why is it necessary to add the potato extract to each tube last, after the water and catechol are already measured?

Why was a different amount of water added to each tube?

E X E R C I S E   5 . 3
Designing an Experiment

Objective

After completing this exercise, you should be able to

1. Design an original experiment to investigate some aspect of enzyme activity.

In Exercise 5.2 you learned a method of measuring the reaction as the enzyme catecholase converts catechol to benzoquinone. In Exercises 5.3 and 5.4 your lab team will design an experiment using this method, perform your experiment, and present and interpret your results. You may want to review Exercise 5.1 to help you decide on an independent variable for your investigation.

The following materials will be supplied for your group.

50 mL potato extract	30 Parafilm squares
2 1-mL pipets	30 test tubes
3 5-mL pipets	test tube rack
pi-pump for 1-mL pipet	50 mL of 0.05% catechol
pi-pump for 5-mL pipet	bottle of distilled water
wax marker	

Your instructor will be able to tell you what additional materials will be available.

Describe your experiment below.

Question or Hypothesis

Dependent Variable

Independent Variable

Explain why you think this independent variable will affect catecholase activity.

Control Treatment(s)

Replication

Brief Explanation of Experiment

Predictions

What results would support your hypothesis? What results would prove it false?

Method

Include the levels of treatment you plan to use. It might be helpful to make a table like Table 5.1, showing the contents of each reaction tube.

Design a Table to Collect Your Data

List Any Additional Materials You Will Require

EXERCISE 5.4

Performing the Experiment and Interpreting the Results

Objectives

After completing this exercise, your should be able to

1. Perform the experiment your lab team designed.
2. Present and interpret the results of your experiment.

Before you do the experiment, be sure that everyone on your lab team understands the techniques that will be used. You may want to divide up the tasks before you begin work. Since it is important to measure the volumes of reactants accurately, you may want to ask your instructor to review pipet use with you.

Be thorough in collecting data. Don't just write down numbers; record what they mean as well. Don't rely on your memory for information that you will need when reporting on your experiment later! If you have any questions, doubts, or problems during the experiment, be sure to write them down, too.

Results

Before you begin to prepare your results for presentation, decide on the best format to use. Remember, you want to give the reader a clear, concise picture of what your experiment showed. Refer to the data presentation exercise in Appendix A (Tools for Scientific Inquiry) for help. If you are drawing graphs, use graph paper. Complete your tables and/or graphs before attempting to interpret your results.

Write a few sentences *describing* the results (don't explain why you got these results or draw conclusions yet).

Discussion

Look back at the hypothesis or question you posed in this experiment. Look at the graphs or tables of your data. Do your results support your hypothesis or prove it to be false? Use your data to support your answer.

Did your results correspond to the prediction you made? If not, explain how your results are different from your expectations and why this might have occurred.

Describe how your data are supported by information from other sources (for example, textbooks or other lab teams working on a similar problem).

If you had any problems with the procedure or questionable results, explain how they might have influenced your conclusion.

If you had an opportunity to repeat and extend this experiment to make your results more convincing, what would you do?

Summarize the conclusion you have drawn from your results.

Questions for Review

1. A freshly cut potato turns brown when left standing. Why do mashed potatoes stay white?

2. The three graphs in Figure 5.11 represent three different enzymes that are found in a unicellular pond organism. Assuming that these enzymes must function in order for the organism to grow, what water pH is best for this organism? Explain your answer.

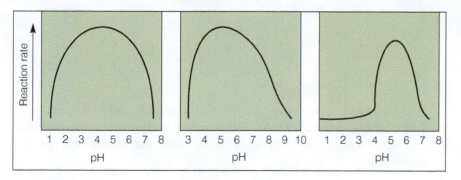

Figure 5.11.
Reaction rate versus pH of water for three enzymes.

3. Lemon juice, which has a pH of about 3, can be sprinkled on freshly cut fruit to keep it from turning brown. Propose a hypothesis to explain this observation.

How could you test your hypothesis?

Propose an alternative hypothesis. (Hint: Lemon juice is a complex substance.)

4. You hypothesize that reaction rate is proportional to enzyme concentration and design an experiment to test your hypothesis. Fill in the blanks below to complete the design.

	Water	**Substrate**	**Enzyme**
Tube A	20 drops	_____	_____
Tube B	15 drops	10 drops	5 drops
Tube C	_____	10 drops	_____
Tube D	_____	_____	20 drops

Explain your answers:
Water:

Substrate:

Enzyme:

5. A student team is studying the effect of substrate concentration on the rate of an enzymatic reaction. They supply each reaction mixture with the same amount of enzyme. Predict what the results will be and explain why.

Cellular Respiration

Introduction

In the process of photosynthesis, plants capture light energy and use it to make the monosaccharide glucose. The energy from the light is packaged into the chemical bonds of the glucose molecule. This glucose can be converted into a number of other molecules, which are either used or stored in the plant. For example, glucose made in the leaves is changed into sucrose, a disaccharide, before it is transported to other parts of the plant. Glucose may also be stored as starch, which is a long chain of glucose molecules. When animals eat plants, the starch and sucrose can be changed back into glucose for use in metabolism. Animals also store excess glucose in long chain molecules (glycogen) or by converting it into fat. Starch and glycogen can in turn be reconverted into glucose when energy is needed by the plant or animal.

The energy contained in the glucose molecule cannot be used directly by the cell. That energy must first be repackaged in ATP (adenosine triphosphate), the molecule that provides the energy for most cellular work. The process of **cellular respiration** transfers the energy stored in glucose bonds to bonds in ATP so that it can be used more easily by the cell. Each glucose molecule can generate as many as 38 ATP molecules through cellular respiration.

In this lab topic we will consider two methods by which cells make ATP using the energy that comes from breaking down the chemical bonds in glucose. These two methods, called **metabolic pathways,** are alcoholic fermentation and aerobic respiration.

Outline

Exercise 6.1: Alcoholic Fermentation

Exercise 6.2: Aerobic Respiration

Exercise 6.3: Designing an Experiment

Exercise 6.4: Performing the Experiment and Interpreting the Results

EXERCISE 6.1
Alcoholic Fermentation

Objectives

After completing this exercise, you should be able to

1. Describe alcoholic fermentation, naming the substrate and products.
2. Explain how alcoholic fermentation can be measured.
3. List factors that can affect the rate of alcoholic fermentation.

Alcoholic fermentation is a metabolic pathway used primarily by yeasts and some bacteria. In fermentation, glucose is broken down into ethyl alcohol (ethanol) and carbon dioxide (Figure 6.1). In the process, some of the energy that had been stored in the glucose bonds is used to form high energy bonds in ATP.

Figure 6.1.
Substrate and products of alcoholic fermentation.

The arrows in Figure 6.1 indicate that a series of enzyme-catalyzed reactions is needed to complete the conversion of glucose (the substrate) into alcohol, carbon dioxide, and ATP (the products). Therefore, any of the factors that affect enzyme activity can affect the rate at which fermentation occurs.

Name some factors that affect enzyme activity.

Although glucose is the substrate at the beginning of the pathway, keep in mind that other molecules can be changed into glucose and used in cellular respiration. For example, starch or glycogen (storage molecules) can be broken down into their many component glucose molecules. Sucrose (table sugar) can be broken down into its two component sugars, glucose and fructose. The fructose can then be converted to glucose by another enzymatic process.

In the procedure you will use, fermentation is performed by yeast, a single-celled fungus. That is, yeast contains the cellular machinery, including the **enzymes,** which is capable of breaking down glucose by alcoholic fermentation. Corn syrup, which contains sucrose and fructose, will be used as the **substrate.** You will measure the rate of alcoholic fermentation by collecting carbon dioxide (CO_2), which is one of the **products,** at intervals after fermentation has begun.

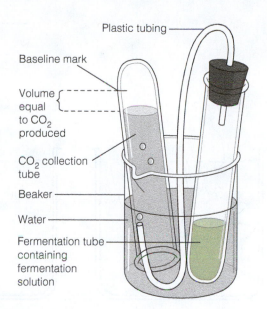

Plastic tubing

Baseline mark

Volume equal to CO_2 produced

CO_2 collection tube

Beaker

Water

Fermentation tube containing fermentation solution

Figure 6.2.
Apparatus for measuring CO_2 production in alcoholic fermentation.

Figure 6.2 shows the setup you will use to collect the CO_2. Fermentation will take place in the test tube on the right, which contains the yeast and corn syrup, the fermentation solution. The test tube is capped with a rubber stopper to prevent CO_2 from escaping. Plastic tubing leads from the fermentation tube to the CO_2 collection tube, which is upside down and contains water. As CO_2 is produced by fermentation, it goes through the tubing into the collection test tube, where it displaces the water. The displacement (in mm) is recorded as a measure of the amount of CO_2 produced.

Procedure

1. You will need to have three fermentation setups, so you should have six large test tubes, three pieces of plastic tubing that have been inserted into rubber stoppers, and three beakers (400 or 600 mL). Using a wax pencil, label three of the test tubes 1, 2, and 3 and set them aside. Assemble the setups one at a time following steps 2–7 (see Figure 6.3 on the next page).

2. Fill a tub or sink with *hot* water (50°–60°C).

3. Insert the end of the plastic tubing into one of the test tubes. This tube will be the CO_2 collection tube. Submerge the collection tube and plastic tubing in the tub of hot water (Figure 6.3a).

4. Submerge the beaker. Place the collection tube in the beaker in an inverted position (Figure 6.3b).

5. Bring the beaker out of the water. One end of the plastic tubing should still be inserted in the collection tube. Hold up the other end of the tubing (the one with the rubber stopper on it) so that the water won't be siphoned out.

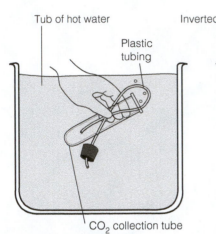

Tub of hot water

Plastic tubing

Inverted collection tube

CO_2 collection tube

a.

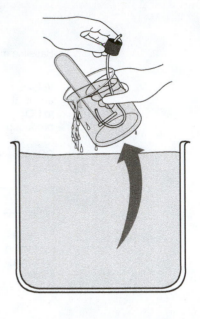

b.

c.

Figure 6.3.
Setting up the CO_2 collection tube.

6. Pour some water out of the beaker so that the water level is at least 1 cm from the top of the beaker (Figure 6.3c).

7. Check the tubing for kinks. If the CO_2 can't get through the tubing, you'll have to start over.

❋ **Assemble all three setups before proceeding to mix the fermentation solutions.**

8. Mix the fermentation solutions for Tubes 1, 2, and 3 according to Table 6.1.

Table 6.1
Contents of Fermentation Tubes

	Fermentation Tube (volume in mL)		
	1	2	3
Water	4	3	1
Yeast suspension	0	1	3
Corn syrup	3	3	3

9. Swirl each test tube gently to mix the reactants. Place one test tube in each beaker.

10. Put the rubber stoppers in the fermentation tubes. This will force most of the water out of the tubing.

11. After the air bubbles from inserting the stopper have cleared the tubing (half a minute to a minute), mark the water level on each collection tube with a wax pencil. This marks the baseline for your experiment.

 If the water level is all the way to the top of the collection tube, where the tube is curved, you should wait until it has descended to the part of the tube where the sides are straight before you mark the level.

12. At 5-minute intervals, measure (in mm) the distance from the baseline mark to the water level. Continue taking data for at least 20 minutes. Record your data in Table 6.2.

Table 6.2
Results of Fermentation Experiment

Tube	Minutes						
	0	5	10	15	20	25	30
1 (3 mL corn syrup)							
2 (3 mL corn syrup + 1 mL yeast suspension)							
3 (3 mL corn syrup + 3 mL yeast suspension)							

While you are waiting to collect data, look again at Table 6.1 to see how the experiment was designed.

What is the independent variable?

What is the dependent variable?

What hypothesis is being tested?

Predict the outcome of the experiment in terms of your hypothesis. Explain what results would support your hypothesis.

The instructor will collect data from all the student teams. Average the data and, in Figure 6.4, plot the average water displacement for each interval against time. This shows the reaction rate (CO_2 produced/time). The steeper the slope of the line, the faster the reaction rate. Be sure to label the axes completely.

Figure 6.4.
Rate of CO_2 production by yeast fermentation.

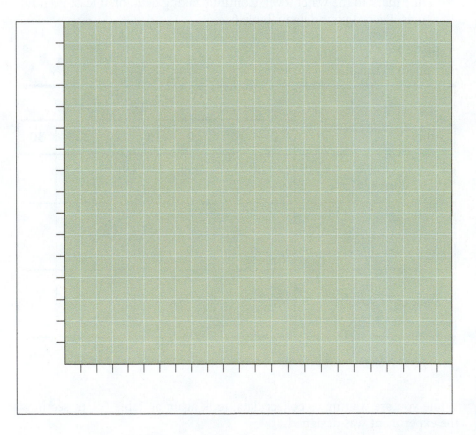

Was your hypothesis proven false or supported by the results? Use data to support your answer.

Which fermentation tube was the control?

Why were different amounts of water added to each fermentation solution?

What are some other independent variables that could affect alcoholic fermentation?

E X E R C I S E   6 . 2
Aerobic Respiration

Objectives

After completing this exercise, you should be able to

1. Name the three phases of aerobic respiration.
2. Explain how aerobic respiration can be measured using DCPIP and the succinate → fumarate reaction.
3. List factors that can affect the rate of aerobic respiration.

Most organisms, including humans, produce most of their ATP by using the process of **aerobic respiration** rather than by fermentation. Aerobic respiration begins with glucose, the same substrate that is used in fermentation. However, aerobic respiration extracts much more of the energy stored in the bonds of each glucose molecule. A single glucose molecule can yield as many as 38 ATP molecules through aerobic respiration, compared with a yield of two ATP per glucose molecule from alcoholic fermentation.

Like fermentation, aerobic respiration begins with the metabolic pathway (series of enzyme-catalyzed reactions) called **glycolysis.** In aerobic respiration, however, pyruvate (which is the product of glycolysis) enters another metabolic pathway called the **Krebs cycle.** Some of the reactions in the Krebs cycle release protons and electrons, which then enter the **electron transport system.** Most of the ATP molecules that are gained in aerobic respiration are produced by the electron transport system.

Glycolysis takes place in the cytoplasm, but the Krebs cycle and electron transport occur within a specific organelle, the **mitochondrion.** Mitochondria are found in all eukaryotic organisms (organisms whose cells have a true nucleus and membrane-bound organelles). You will use a suspension made from ground-up lima beans to study aerobic respiration. The suspension contains mitochondria, which, under our experimental conditions, will continue to carry out aerobic respiration as if they were in intact cells. Sucrose has been added to the mitochondrial suspension as a source of glucose for respiration, just as corn syrup was used in alcoholic fermentation.

The rate of aerobic respiration can be measured by studying the activity of one particular enzyme in the Krebs cycle. This enzyme catalyzes a reaction that converts succinate, which is the substrate for the reaction, to fumarate (Figure 6.5).

Figure 6.5.
Breakdown of glucose by glycolysis and the Krebs cycle. The enzyme-catalyzed conversion of succinate to fumarate is one step in the Krebs cycle.

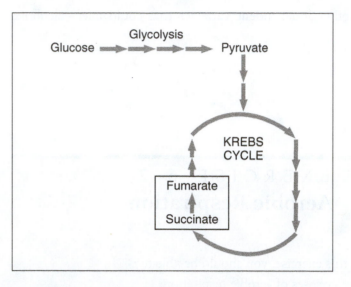

As the enzymatic reaction occurs, protons and electrons are released from the succinate molecule. In the living mitochondrion, these protons and electrons are carried to the electron transport system and used to make ATP. In order to investigate aerobic respiration, you will add a blue compound, DCPIP (di-chlorophenol-indophenol) to the reaction mixture. DCPIP intercepts the protons and electrons before they get to the electron transport system. When a DCPIP molecule picks up a proton and electron, its blue color disappears and it becomes colorless. We can use this color change as a measure of the enzymatic reaction rate, which in turn is a measure of the rate of aerobic respiration. As succinate is converted to fumarate (and as aerobic respiration proceeds), the solution in the test tube gradually turns from blue to colorless. The reaction is summarized in Figure 6.6.

Figure 6.6.
Color change of DCPIP caused by the conversion of succinate to fumarate.

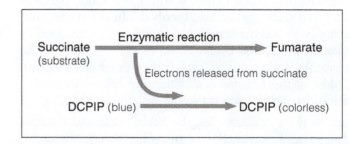

The color change that DCPIP undergoes thus provides a way to visualize aerobic respiration as it occurs. We now need a method of measuring how much color change occurs. An instrument called a spectrophotometer can be used. A common model (the one referred to in this lab topic) is the Spectronic 20 (Spec 20). You may have a different model in your laboratory.

If spectrophotometers are not available, you can use a color chart to determine how much color change has taken place. Compare your samples with Plate 2 and record the number for the color intensity that matches the sample most closely.

As shown in Figure 6.7, the Spec 20 shines light through a sample of reactants in a tube and measures the amount of light that penetrates through the tube. In Lab Topic 5 (Enzymes), you recorded data from the absorbance scale: The data showed how much light had been absorbed by each sample. In this experiment you will record data from the % transmittance scale.

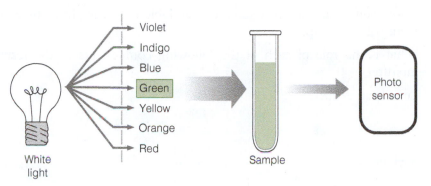

Figure 6.7.
Diagrammatic representation of how a Spec 20 works. Like a prism, a spectrophotometer divides light into its component wavelengths.

Your data will show how much light has been transmitted by the sample. The more intense the color of the sample, the less light it transmits. Percent transmittance measures how much of the blue color has disappeared. The blue DCPIP becomes colorless as aerobic respiration proceeds, so the higher the transmittance reading the more mitochondrial activity has been observed.

On the axes in Figure 6.8, sketch the relationship between % transmittance or color intensity and mitochondrial activity (aerobic respiration). Label both axes.

Figure 6.8.
Relationship of % transmittance or color intensity to mitochondrial activity.

Table 6.3

Reaction Mixtures for Mitochondrial Activity Experiment

	Tube (volume in mL)		
	1	2	3
Buffer	4.4	4.3	4.2
DCPIP	0.3	0.3	0.3
Mitochondrial suspension	0.3	0.3	0.3
Succinate	0	0.1	0.2

Before you begin the procedure, look at Table 6.3 to see how the experiment will be set up.

What is the role of each of the following components of the reaction mixture?

Lima bean extract:

Succinate:

DCPIP:

Buffer:

Why should there be a different amount of buffer in each tube?

What is the independent variable?

What is the dependent variable?

What hypothesis is being tested?

Predict the outcome of the experiment in terms of your hypothesis. Explain what results would support your hypothesis.

Procedure

Zeroing the Spec 20 (skip steps 1–7 if you are using the color chart)

When you look at the mitochondrial suspension, you will notice that it is cloudy. So even though it is colorless, transmittance of light through the sample will already be less than 100% before any DCPIP is added. If you are using a Spec 20, you will therefore prepare a "blank" tube containing mitochondrial suspension but no DCPIP and use that tube to set the transmittance at 100%. This is called zeroing the machine.

1. Get a clean test tube (either a small test tube or a special Spec 20 tube) and use a wax pencil to label it B. When you put the tube into the sample holder, it should always be inserted with the B facing front.

2. Measure 4.6 mL buffer, 0.3 mL mitochondrial suspension, and 0.1 mL succinate into Tube B.

3. Cover the Tube B tightly with Parafilm and invert it to mix the reactants thoroughly.

4. Set the wavelength of the Spec 20 to 600 nm (Figure 6.9).

5. Use the knob on the left to set % transmittance (top scale) to 0.

6. Wipe the blank tube (B) with a cleaning tissue and insert it into the sample holder.

7. Use the knob on the right to set % transmittance to 100%.

8. Set Tube B aside. You will need to rezero the Spec 20 later.

The Spec 20 is now zeroed and ready to measure sample tubes for the experiment.

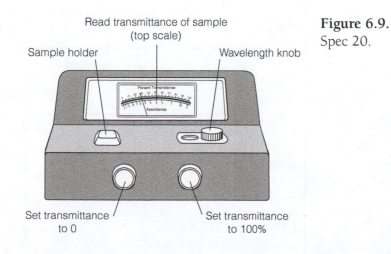

Figure 6.9.
Spec 20.

Preparing Reaction Mixtures for the Experiment

9. Get three Spec 20 tubes or small test tubes and use a wax pencil to label them 1, 2, and 3.

10. Measure the buffer, DCPIP, and mitochondrial suspension into Tubes 1, 2, and 3 as specified in Table 6.3, but it is important not to add the succinate until the other ingredients are in the tubes.

11. As quickly as possible, add the succinate to each tube.

12. Cover each tube tightly with Parafilm and invert to mix.

Spec 20 Method

13. As you complete each tube, insert it into the sample holder of the Spec 20 with its number facing to the front and record the % transmittance in Table 6.4 at time 0.

14. After 5 minutes, use the blank to rezero the Spec 20 (steps 5–7). Invert each tube to remix its contents and read and record the % transmittance.

15. Continue to record the % transmittance in the tubes at 5-minute intervals for at least 20 minutes or until Tube 3 is nearly colorless (% transmittance is near 100%). You should rezero the Spec 20 before each reading and always insert the tubes with the number facing the front.

Color Chart Method

13. Compare each tube to the right chart on Color Plate 1. Observe the intensities of the colors. In Table 6.4, record the number of the color that most closely matches the color intensity in each tube at time 0. At the beginning of the experiment all the tubes should be the same color.

Table 6.4
Results of Mitochondrial Activity Experiment
(% transmittance or color intensity)

Tube	Minutes						
	0	5	10	15	20	25	30
1 (0.3 mL mitochondrial suspension)							
2 (0.3 mL mitochondrial suspension + 0.1 mL succinate)							
3 (0.3 mL mitochondrial suspension + 0.2 mL succinate)							

* Look for a similarity in color intensity. The exact color does not have to match.

14. After 5 minutes, shake each tube gently to mix its contents and compare it to Color Plate 1. In Table 6.4, record the number of the color that most closely matches the color intensity in each tube.

15. Continue to record the color in the tubes at 5-minute intervals for 20 minutes or until Tube 3 is colorless (0).

The instructor will collect data from all the student teams. Average the data and plot the average % transmittance (or color intensity) for each interval against time in Figure 6.10. This shows the reaction rate for aerobic respiration. The steeper the slope of the line, the faster the reaction rate. Be sure to label the axes completely.

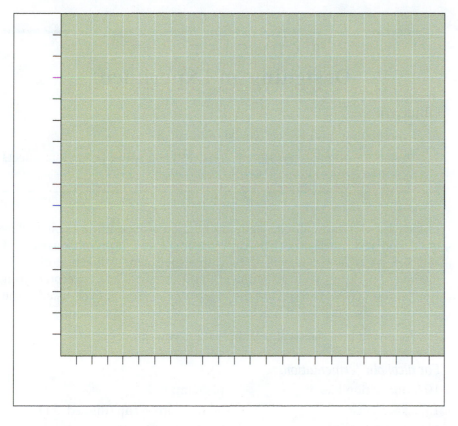

Figure 6.10.
Rate of aerobic respiration in mitochondria.

Was the hypothesis proven false or supported by the results? Use data to support your answer.

Which reaction tube was the control?

Why should the succinate be added to the reaction tubes last?

What are some other independent variables that could affect aerobic respiration?

✳ When you design your own experiment to investigate mitochondrial activity using the Spec 20, be sure to include a blank tube for zeroing the instrument.

E X E R C I S E   6 . 3
Designing an Experiment

Objective

After completing this exercise, you should be able to

1. Design an original experiment to investigate some factor that affects alcoholic fermentation or aerobic respiration.

In Exercises 6.1 and 6.2, you learned a method of measuring alcoholic fermentation and a method of measuring aerobic respiration. In Exercises 6.3 and 6.4, your lab team will design an experiment using one of these methods, perform your experiment, and present and interpret your results. You may want to review the independent variables you listed at the end of Exercises 6.1 and 6.2 before you design your experiment.

The following materials will be supplied for your group.

For alcoholic fermentation:

10 fermentation kits	pi-pump
1 package yeast	50 mL corn syrup (diluted 1:1)
150-mL beaker	bottle of water
glass stir rod	wax marker
3 5-mL pipets	ruler

For aerobic respiration:

20 test tubes	pi-pump for 5-mL pipet
test tube rack	20 Parafilm squares
3 1-mL pipets	25 mL DCPIP
1 5-mL pipet	25 mL succinate
pi-pump for 1-mL pipet	100 mL buffer solution

Your instructor will be able to tell you what additional materials will be available.

Describe your experiment below.

Question or Hypothesis

Dependent Variable

Independent Variable

Explain why you think this independent variable will affect alcoholic fermentation or aerobic respiration.

Control(s)

Replication

Brief Explanation of Experiment

Predictions

What results would support your hypothesis? What results would prove your hypothesis false?

Method

Include the levels of treatment you plan to use. It might be helpful to make a table showing the contents of each reaction vessel.

Design a Table to Collect Your Data

List Any Additional Materials You Will Require

EXERCISE 6.4
Performing the Experiment and Interpreting the Results

Objectives

After completing this exercise, you should be able to

1. Perform the experiment your lab team designed.
2. Present and interpret the results of your experiment.

Before you do the experiment, be sure that everyone on your lab team understands the techniques that will be used. You may want to divide up the tasks before you begin work. Since it is important to measure the volumes of reactants accurately, you may want to ask your instructor to review pipet use with you.

Be thorough in collecting data. Don't just write down numbers; record what they mean as well. Don't rely on your memory for information you will need when reporting on your experiment later! If you have any questions, doubts, or problems during the experiment, be sure to write them down, too.

Results

Before you begin to prepare your results for presentation, decide on the best format to use. Remember, you want to give the reader a clear, concise picture of what your experiment showed. Refer to the data presentation section of Appendix A (Tools for Scientific Inquiry) for help. If you are drawing graphs, use graph paper. Complete your tables and/or graphs before attempting to interpret your results.

Write a few sentences *describing* the results (don't explain why you got these results or draw conclusions yet).

Discussion

Look back at the hypothesis or question you posed in this experiment. Look at the graphs or tables of your data. Do your results support your hypothesis or prove it false? Use your data to support your answer.

Did your results correspond to the prediction you made? If not, explain how your results are different from your expectations and why this might have occurred.

Describe how your data are supported by information from other sources (for example, textbooks or other lab teams working on a similar problem).

If you had any problems with the procedure or questionable results, explain how they might have influenced your conclusion.

If you had an opportunity to repeat and extend this experiment to make your results more convincing, what would you do?

Summarize the conclusion you have drawn from your results.

Questions for Review

1. What is the importance of cellular respiration to living organisms?

2. Yeast is used to make bread rise. How might this work?

3. Which of the following tubes would you expect to produce the highest initial rate of CO_2 production? Explain your answer.

    Tube 1: 5 mL corn syrup, 10 mL yeast, 5 mL water

    Tube 2: 10 mL corn syrup, 10 mL yeast, 0 mL water

    Tube 3: 0 mL corn syrup, 10 mL yeast, 10 mL water

4. Suppose you do an experiment on aerobic respiration using lima bean mitochondria as described in Exercise 6.2 and the results show a final % transmittance reading of 42% in Tube 1 and 78% in Tube 2. Both tubes had an initial reading of 15%. In which tube did the greatest amount of aerobic respiration occur? Explain your answer in terms of the changes that take place in DCPIP.

Acknowledgment

The procedure used to assay mitochondrial activity was based on a procedure from Succinic Acid Dehydrogenase Activity of Plant Mitochondria in F. Witham, D. Blaydes, and R. Devlin, *Exercises in Plant Physiology*, Boston: Prindle, Weber & Schmidt, 1971.

Photosynthesis

Introduction

The process of photosynthesis can be summarized in the equation below.

$$\text{Chloroplasts}$$
$$6\ CO_2 + 12\ H_2O \longrightarrow C_6H_{12}O_6 + 6\ H_2O + 6\ O_2$$
$$\text{Light}$$

This simple equation conceals a multitude of complex processes, which are divided into two phases. In the first phase, known as the light-dependent reactions, the pigment chlorophyll absorbs light energy, which is then invested in the high-energy chemical bonds of ATP and NADPH molecules. The second phase of photosynthesis, known as carbon fixation, does not require light, since the energy has already been captured. In this light-independent phase, ATP and NADPH, along with CO_2, are used to make the carbohydrate glucose, which is a six-carbon monosaccharide (see Lab Topic 3, Macromolecules). So the result of photosynthesis is that a small amount of light energy is absorbed and then stored in the chemical bonds of glucose. The entire process takes place within the chloroplasts of plants. The chloroplast contains the chlorophyll, the enzymes, and the other molecules that are required for photosynthesis. (The prokaryotic Cyanobacteria, which can also photosynthesize, do not have their photosynthetic apparatus enclosed in an organelle.)

As you learned in Lab Topic 6 (Cellular Respiration), virtually all of the energy used by organisms on the earth, with the exception of a few chemosynthetic organisms, is made available by photosynthesis. In the process of cellular respiration, the energy stored in the chemical bonds of carbohydrates is released and repackaged in ATP, the usable "energy currency" of the cell.

In this lab topic you will learn a method of measuring the rate of photosynthesis. You will then use this method to design your own investigation of photosynthesis.

Outline

Exercise 7.1: Measuring Photosynthetic Rate in Spinach Leaf Disks
Exercise 7.2: Designing an Experiment
Exercise 7.3: Performing the Experiment and Interpreting the Results

E X E R C I S E   7 . 1
Measuring Photosynthetic Rate in Spinach Leaf Disks

Objectives

After completing this exercise, you should be able to

1. Explain how the experiment measures the occurrence of photosynthesis in spinach.
2. Interpret the results of the experiment.

The light-dependent reactions of photosynthesis are summarized in the equation:

Chloroplasts

$$H_2O + ADP + ⓅP + NADP^+ \rightarrow \frac{1}{2}O_2 + ATP + NADPH + H$$

Light

Light energy is used to split water molecules. Some of that energy is captured by adding P (phosphate) to ADP to make ATP and by adding hydrogen to $NADP^+$. O_2 is released as a by-product. The rate of these reactions can be estimated by measuring O_2 production in disks cut from spinach leaves.

Leaf tissue is riddled with gas-filled intercellular spaces (see Figure 7.1a), so they float when they are placed in solution (Figure 7.1b). But when leaf disks in a solution are subjected to a vacuum, the gases in the leaves are pulled out and the spaces are filled by the liquid. Since fluids are heavier than gas, replacing the gas causes the leaf disks to sink to the bottom of the flask (Figure 7.1c). Then as O_2 is produced by the light reactions of photosynthesis, it diffuses into the intercellular spaces and replaces the liquid with gas. When enough O_2 has accumulated, each leaf disk will regain its buoyancy and turn on edge or float to the surface (Figure 7.1d). You will use this technique as a method of measuring photosynthesis. Sodium bicarbonate ($NaHCO_3$), which supplies CO_2 for photosynthesis, will be the solution used to infiltrate the leaf disks.

Procedure

1. Attach the lamp to the support stand so that the lamp is approximately 25 cm from the base of the stand.
2. Fill a large (1- or 2-L) beaker with cold water to act as a heat filter for the dish you will place under the lamp. The lamps produce a lot of heat, which can affect the rate of photosynthesis. Set the beaker aside for the moment.

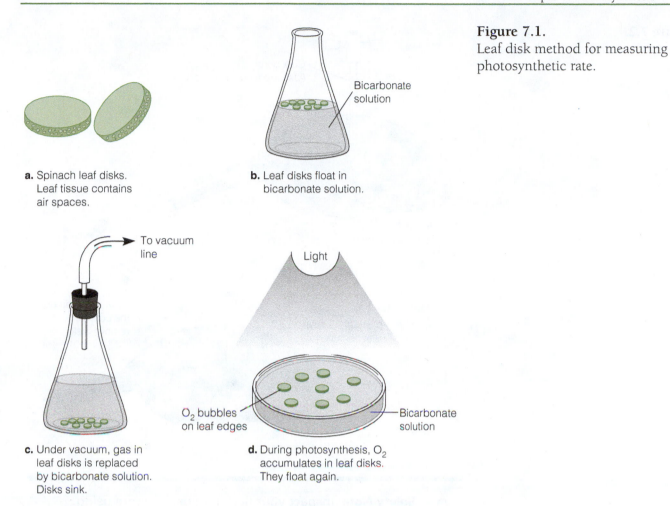

Figure 7.1.
Leaf disk method for measuring photosynthetic rate.

a. Spinach leaf disks. Leaf tissue contains air spaces.

Bicarbonate solution

b. Leaf disks float in bicarbonate solution.

To vacuum line

Light

O_2 bubbles on leaf edges

Bicarbonate solution

c. Under vacuum, gas in leaf disks is replaced by bicarbonate solution. Disks sink.

d. During photosynthesis, O_2 accumulates in leaf disks. They float again.

3. Pour 0.2% $NaHCO_3$ solution into three petri dishes so they are about 2/3 full. Pour approximately 100 mL of 0.2% $NaHCO_3$ solution into a 250-mL flask. The exact amount is not important, so you can use the markings on the flask to measure.

4. Get several spinach leaves and cut 40–50 disks with a cork borer or other circular cutting instrument. (See Figure 7.2 on the next page.) Do not include the large veins. Use a paper towel or Styrofoam board to cut on. As the disks are cut, put them in the flask with the bicarbonate solution.

Stack three or four leaves together so you can cut through them all at the same time. But don't cut so many leaves that the edges of the disks are ragged.

5. Use a water aspirator or other vacuum source, as directed by your instructor, to sink the disks.

Figure 7.2.
Cutting disks from spinach leaves.

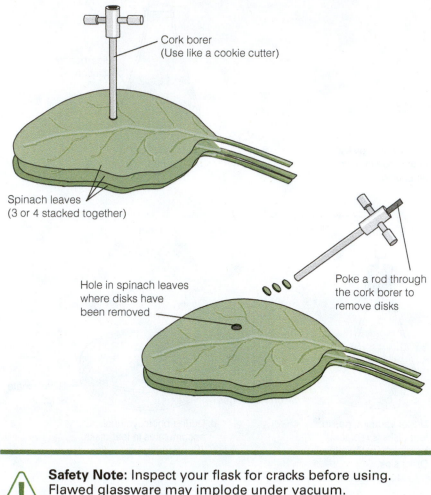

Cork borer
(Use like a cookie cutter)

Spinach leaves
(3 or 4 stacked together)

Hole in spinach leaves
where disks have
been removed

Poke a rod through
the cork borer to
remove disks

Safety Note: Inspect your flask for cracks before using.
Flawed glassware may implode under vacuum.
Wear safety goggles while using the aspirator.

a. Attach the vacuum tubing to the sidearm of the flask or the glass tube in the rubber stopper. Put the rubber stopper firmly in the mouth of the flask and press the tape securely over the hole in the stopper.

Hold the flask steady while you apply the vacuum.

b. Turn on the water faucets all the way. After several seconds, you should see bubbles coming out of the edges of the disks. Leave the disks under vacuum for approximately 15–20 seconds; then release the vacuum by peeling back the tape.

c. Swirl the flask and wait to see whether the disks sink. Remember, the leaves will continue to float as long as they are under vacuum.

d. You will probably have to apply the vacuum two or three times. It is not necessary to sink all the disks. You will damage the tissue if you overaspirate.

✳   **Keep the disks away from bright light!**

---

6. Pour the contents of the flask into a large dish. *Discard any disks that are floating.* With forceps, gently transfer 10–15 disks to each petri dish. Put the lids on the petri dishes. Put one dish in a dark place (drawer or cabinet); put another one under the lamp. Set the large beaker of water on top of the petri dish; then turn on the lamp. Place the third dish on a bench where it will receive only room light.

7. Wait 20 minutes for photosynthesis to occur. While the experiment is running, answer the questions below and work on Exercise 7.2.

8. After 20 minutes, count the number of disks that are either floating or turned on edge in each petri dish. Record your data in Table 7.1.

Table 7.1
Results of Subjecting Spinach Leaf
Disks to Different Light Conditions

Light	# Disks Floating	% Disks Floating
Dark		
Room light		
Under lamp		

What hypothesis is being tested with this experiment?

Predict what the results will be if your hypothesis is supported.

What is the independent variable in this experiment?

What are some other independent variables that could affect photosynthesis?

What is the dependent variable?

What is the control treatment?

Was your hypothesis supported or proven false by the results? Explain.

E X E R C I S E 7 . 2
Designing an Experiment

Objective

After completing this exercise, you should be able to

1. Design an original experiment to investigate some factor that affects the production of oxygen in photosynthesis.

In Exercise 7.1 you learned a method of measuring photosynthetic rate in spinach leaf disks. In Exercises 7.2 and 7.3 your lab team will design an experiment using this method, perform your experiment, and present and interpret your results. Review the factors that could affect photosynthesis that you listed in Exercise 7.1 to get some ideas for possible independent variables.

The basic materials that are needed to perform an investigation will be supplied for you. Your instructor will tell you what additional materials will be available.

fresh spinach	6 petri dishes
0.2% sodium bicarbonate (1 L)	2 reflector lamps
250-mL flask with 2-hole rubber stopper	2 support stands for lamps
vacuum source	Two 1-L or 2-L beakers
#3 cork borer or other cutting instrument	culture dish
glass stir rod	forceps
	cutting board

Describe your experiment below.

Question or Hypothesis

Dependent Variable

Independent Variable

Explain why you think this independent variable will affect photo-synthesis.

Control Treatment(s)

Replication

Brief Explanation of Experiment

Predictions

What results would support your hypothesis? What results would prove your hypothesis false?

Method

Include the levels of treatment you plan to use.

Design a Table to Collect Your Data

List Any Additional Materials You Will Require

E X E R C I S E   7 . 3
Performing the Experiment and Interpreting the Results

Objectives

After completing this exercise, you should be able to

1. Perform the experiment your lab team designed.
2. Present and interpret the results of your experiment.

Before you do the experiment, be sure that everyone on your lab team understands the techniques that will be used. You may want to divide up the tasks before you begin work.

Be thorough in collecting data. Don't just write down numbers; record what they mean as well. Don't rely on your memory for information that you will need when reporting on your experiment later! If you have any questions, doubts, or problems during the experiment, be sure to write them down, too.

Results

Before you begin to prepare your results for presentation, decide on the best format to use. Remember, you want to give the reader a clear, concise picture of what your experiment showed. Refer to the data presentation section of Appendix A (Tools for Scientific Inquiry) for help. If you are drawing graphs, use graph paper. Complete your tables and/or graphs before attempting to interpret your results.

Write a few sentences *describing* the results (don't explain why you got these results or draw conclusions yet).

Discussion

Look back at the hypothesis or question you posed in this experiment. Look at the graphs or tables of your data. Do your results support your hypothesis or prove it to be false? Explain your answer, using your data for support.

Did your results correspond to the prediction you made? If not, explain how your results are different from your expectations and why this might have occurred.

Describe how your data are supported by information from other sources (for example, textbooks or other lab teams working on a similar problem).

If you had any problems with the procedure or questionable results, explain how they might have influenced your conclusion.

If you had an opportunity to repeat and extend this experiment to make your results more convincing, what would you do?

Summarize the conclusion you have drawn from your results.

Questions for Review

1. What is the function of each of the following in photosynthesis?
   Light:

CO_2:

Chlorophyll:

2. In the leaf disk method, spinach tissue is infiltrated with bicarbonate solution so that it sinks. Why do the disks float again after being exposed to light?

3. Reflotation of leaf disks by O_2 is one method of measuring photosynthesis. Review the equations for photosynthesis and suggest at least one other method that could be used (assume that you have the means of measuring any substance you choose).

4. A team of students performs an experiment to determine whether different wavelengths of light are equally effective at producing photosynthesis. Their results are shown in Table 7.2. What conclusion would you draw from these results?

Table 7.2
Effects of Wavelength on Photosynthesis

Light	# Disks Floating	% Disks Floating
Dark	1/30	3
White light	25/30	83
Blue light	24/30	80
Yellow light	3/30	10
Green light	2/30	7
Red light	25/30	83

Acknowledgments

The leaf disk technique was modified from Witham, F., D. Blaydes, and R. Devlin. *Experiments in Plant Physiology.* New York: Van Nostrand Reinhold, 1971. Those authors acknowledge the following paper for the original technique: Wickliff, J. L., and R. M. Chasson. "Measurement of Photosynthesis in Plant Tissues Using Bicarbonate Solutions." *Bioscience* 14:32–33, 1964.

Chromosomes and Cell Division

✳ Before you begin the activities in this lab topic, you should be able to

1. Explain the role of DNA in genetics.
2. Explain how the eukaryotic genome is organized.
3. Define the terms chromosome and gamete.
4. Distinguish the general purpose of mitosis from the purpose of meiosis.

Introduction

The genetic information of all organisms is found in deoxyribonucleic acid (DNA), which consists of varying sequences of the four nucleotides adenine, thymine, guanine, and cytosine. Discrete sections of DNA are called genes, and they are the "blueprints" that make organisms. In the cells of eukaryotes such as wheat and humans, DNA is packaged with proteins. Most of the time this DNA-protein complex is found in a threadlike form called chromatin. Recall Lab Topic 4 when you looked at your cheek cells under the microscope. The most prominent feature of the cells was the nucleus, but you needed to use the high-power objective to see it clearly. If the DNA molecules in one of those microscopic nuclei were unwound, attached together, and pulled out like a string, it would be two meters long! It fits inside the nuclei because it is extremely thin. In this lab topic, you will extract DNA from wheat cells to see this famous molecule for yourself.

During cell division, the chromatin is elaborately wound up into the coiled structures called chromosomes. When cells give rise to new cells, there must be a way for each new cell to receive an exact copy of all of the chromosomes. In this lab topic you will examine how the process of **mitosis** accomplishes this.

Genetic information must also be passed from parents to offspring, and in this lab topic you will see how the process of **meiosis** distributes chromosomes into gametes for the purpose of sexual reproduction.

Outline

Exercise 8.1: The Genetic Material

      Activity A: Extracting DNA from Cells

      Activity B: Karyotypes

E X E R C I S E   8 . 1
The Genetic Material

Objectives

After completing this exercise, you should be able to

1. Describe how DNA can be extracted from cells.
2. Explain what a karyotype shows.
3. Define the following terms: homologous chromosomes, centromere, diploid, and haploid.

Activity A: Extracting DNA from Cells

DNA is found in almost every cell of every organism. You've heard about it, you've seen diagrams of its structure, but have you ever actually seen DNA molecules? In this activity you will extract DNA from the embryonic cells of wheat seeds, also known as wheat germ. After softening the cell walls of the wheat germ in warm water, detergent is used to break up the membranes. The DNA is then separated from the rest of the cell contents.

Procedure

1. Weigh out 1 g of wheat germ.
2. Put the wheat germ in a large (50-mL) test tube.
3. Dip warm water from the water bath and measure 20 mL in a graduated cylinder. Pour the water into the test tube with the wheat germ.
4. Using a wooden stick, stir the wheat germ *gently* for 3 minutes.
5. Get a pasteur pipetful of detergent and release it into the test tube with the wheat germ.
6. Stir the wheat germ mixture gently for 5 minutes, being careful not to create any foam.

✳ **Stir slowly to avoid making bubbles!**

7. Measure 15 mL of *cold* alcohol in a graduated cylinder.

8. Tilt the test tube containing wheat germ at a 45-degree angle and *very slowly* pour the alcohol from the graduated cylinder down the side of the wheat germ tube. The alcohol should just trickle down the side and come to rest on the top of the water so that it forms a separate layer.

✳   Pour very slowly, taking care not to let the layers mix.

---

9. Once you have poured in all the alcohol, place the test tube in its rack and do not move it for at least 15 minutes. The DNA will begin precipitating out immediately between the two layers of liquid.

10. After 15 minutes, the DNA should be floating on the top of the test tube. Use the wooden stick to spool it like cotton candy.

Describe what the DNA looks like.

The precipitated DNA in the alcohol layer is only part of the total DNA from the wheat cells. Much of the DNA is still in the water below. You can bring this DNA into the alcohol, where it will precipitate. Tilt the test tube slightly and insert the wooden stick into the yellowish water fraction, then pull the stick up into the alcohol. You should see more DNA come up with it. Be careful not to stir the layers together, though! When it reaches the alcohol, the DNA will precipitate in its stringy form and you can spool it, too.

Most protocols for DNA extraction use an enzyme to digest proteins. That step was omitted in this procedure to get faster results. How would your results be different if you used a protein-digesting enzyme?

Activity B: Karyotypes

Although most of the DNA in a prokaryotic organism such as bacteria is carried on a single chromosome, in eukaryotes (animals, plants, fungi, and protists) the DNA is divided among numerous chromosomes. Each eukaryotic species has a characteristic number of chromosomes, but that number does not indicate the complexity of the organism. For example, redwood trees have 66 chromosomes, gypsy moths have 62, yeast have 32, and catfish have 58. We humans have 46 chromosomes in each body cell. Guppies also have 46 chromosomes, but there are many differences in the genetic "blueprints" that make a guppy and the genetic "blueprints" that make a human.

We can study an organism's chromosomes by making a karyotype of them. To make a karyotype, the nucleus of the cell is ruptured to isolate the chromosomes and then a microscopic digital image is made. A technician uses

Figure 8.1.
Karyotype of human male.

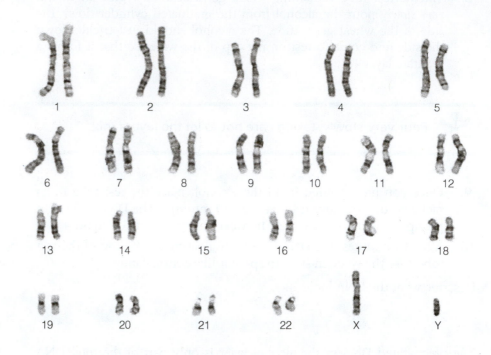

a computer program to arrange the images of the chromosomes in **homologous pairs** (Figure 8.1). The pairs are matched by comparing physical characteristics such as length and staining pattern. Homologous chromosomes have these physical similarities because they are genetically similar: they carry the same genes in the same order.

Figure 8.1 shows a finished karyotype of a human male. In humans there are 22 homologous pairs of autosomes (chromosomes that are not concerned with sex determination) plus a pair of sex chromosomes. Females have a homologous pair of X chromosomes, but the male's X and Y, shown here, don't match.

The karyotype, as well as other pictures of chromosomes that you have seen in books, shows chromosomes from cells that are undergoing mitosis. Thus the chromosomes are distinct, thickly coiled structures rather than the thready chromatin form. You may also be able to see that the chromosomes in Figure 8.1 are doubled, which makes each one look like a slender "X." Before a new nucleus can be made, the DNA in each chromosome must replicate (make an exact copy) so that each new cell will have exactly the same genetic material as the original cell. After replication, the chromosomes remain attached at a place called the **centromere.** The point of attachment is not always in the center of the chromosomes. The position of the centromere is another characteristic that technicians use to match homologous pairs.

How many *homologous pairs* of chromosomes do the following organisms have?

redwood tree—

gypsy moth—

catfish—

guppy—

A cell that contains the correct number of homologous pairs for its species is said to be **diploid.** A cell with only one member of each homologous pair is said to be **haploid.**

Almost all of the cells in our bodies are diploid. What cells are haploid, and why?

Suppose the karyotype shown in Figure 8.1 was taken from a cheek cell. What cell division process would produce a new cheek cell?

Describe what the karyotype of a new cheek cell from this same individual would look like. That is, how many chromosomes would it have? Would it be haploid or diploid?

Suppose we made a karyotype of a gamete from this individual. What would it look like? How many chromosomes would it have? Would it be haploid or diploid?

E X E R C I S E   8 . 2
The Process of Mitosis

Objectives

After completing this exercise, you should be able to

1. Explain the purpose and location of mitosis in organisms.
2. Describe and recognize the phases of mitosis.

The union of a haploid egg with a haploid sperm produces a diploid zygote. The zygote has a nucleus that holds the chromosomes, which in turn contain the entire blueprint to make the organism. From that single cell, the process of **mitosis** produces the many (sometimes trillions of) diploid cells that make up the body of a multicellular organism. Therefore each of the organism's cells contains an exact copy of the same genetic material.

Because cells must constantly be replaced, mitosis also produces new diploid cells throughout the life of an organism. For example, your entire epidermis is replaced every 15 to 30 days, and your body makes approximately 200 billion new red blood cells every day. That's a lot of mitosis!

In the following activities you will examine the steps in mitosis by using models of chromosomes. You will then identify these steps in onion cells.

Activity A: Pop Bead Simulation of Mitosis

You will use pop beads to represent chromosomes and follow them through the processes of cell division. Your instructor will give you a set of "chromosomes" to work with.

Before you begin, take stock of your chromosomes. How many chromosomes are there?

How many homologous pairs of chromosomes?

How can you tell which chromosomes are homologous?

How can you distinguish the members of the homologous pair *from each other*?

What differences between the chromosomes are represented by the different colors? That is, what is different about the maternal and paternal chromosomes?

Procedure

In the space below, sketch your chromosomes. You shouldn't take the time to draw each pop bead, but do be sure that homologous pairs can be identified in your drawing. An example is shown below.

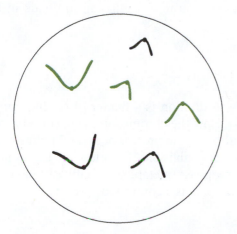

Interphase

Before cell division begins, each chromosome must replicate. This event happens while the chromosomes are still in their long, stringy, uncoiled state, so when interphase is viewed under the microscope, the chromosomes are not even identifiable.

Using more pop beads, make an exact copy of each of your chromosomes. At this point the chromosome is doubled, with an attachment point called the centromere. Each strand of the doubled chromosome is called a chromatid. Because they are identical, chromatids that are attached to each other are called sister chromatids. Sketch the chromosomes below.

How many chromosomes are there now? (As long as the chromatids are attached, the structure is considered one chromosome.)

How many homologous pairs of chromosomes?

How many chromatids?

Besides replication of the chromosomes, what else should happen during interphase? (Hint: The entire cell is going to divide in half, not just the nucleus.)

After interphase, cell division is a matter of sorting out the cell contents into two new cells. How the chromosomes are sorted is of special interest. By taking a "snapshot" of the process at several points, we can understand the mechanism of this division. Traditionally, the "snapshot" phases of mitosis are called prophase, metaphase, anaphase, and telophase, but keep in mind that cell division is a continuous process.

Prophase
It is during prophase that the chromosomes coil more tightly and become visible as distinct structures. Since we already modeled short, thick chromosomes during interphase, there are no changes to our pop bead model.

By the end of prophase, the nuclear envelope disintegrates, enabling the chromosomes to be snared by protein fibers that have grown out of two centrosomes that formed in the cytoplasm. Together these protein fibers form the spindle, an apparatus that will eventually allow the sister chromatids to separate from each other.

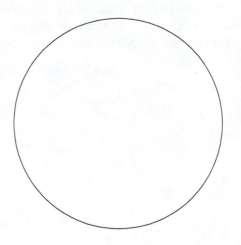

Metaphase
By metaphase the two centrosomes have moved to opposite ends of the cell with their protein fibers attached, and many of these protein fibers are also attached to the centromere region of a chromosome. One sister chromatid of each chromosome is connected to each pole by this spindle apparatus.

The distinctive characteristic of metaphase is the alignment of the chromosomes in a plane through the center of the cell. (Only the centromeres are aligned.) Place your chromosome models in this position and sketch metaphase in the circle below.

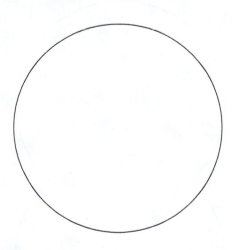

Based on what you know about the outcome of mitosis, can you guess what's going to happen next?

Anaphase

During anaphase the centromeres joining the sister chromatids separate and the chromatids—now called chromosomes—move along the spindle fibers to opposite poles. The number of chromosomes in the cell is briefly double as the cell is prepared for the final phase of division.

Place your chromosome models in the anaphase positions and sketch them in the circle below.

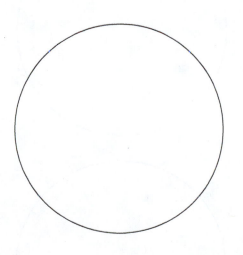

Telophase and Cytokinesis

Now that the chromatids have separated, the division of chromosomes is complete. All that remains is to reconstruct the nuclear envelopes and divide the cytoplasm between the two new cells. These processes will be finished by the end of telophase, and the chromosomes will resume their typical long, stringy form.

Place your chromosomes in the telophase positions and sketch them below.

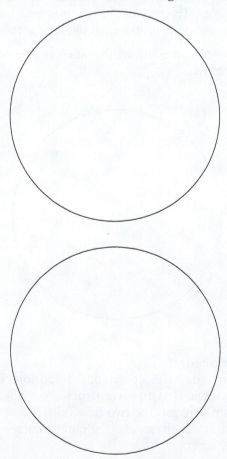

As the chromosomes go through telophase, the division of the cytoplasm, called **cytokinesis**, is also completed. The daughter cells separate from each other. Draw the chromosomes in the daughter cells below.

How many chromosomes are there in each new cell?

How many homologous pairs of chromosomes in each new cell?

What is the overall result of mitosis?

After seeing the phases of mitosis, you may be wondering why cell division is such an elaborate process. Once interphase is complete, why not cut to the chase—telophase? Mitosis has to be carefully choreographed so that each daughter cell has the same number and type of chromosomes—that is, so the division of chromosomes is absolutely equal. Did you ever share a bag of M&Ms with a sibling or friend by carefully counting out and dividing one color at a time? The purpose was to be sure that the distribution was exactly equal, and mitosis has the same purpose regarding chromosomes. On the other hand, you were probably willing to divide a Snickers bar down the middle, without bothering to count every last peanut. You made the assumption that both halves were more or less equal. That's how the cytoplasm and included organelles are divided.

Figure 8.2.
Onion root tip.

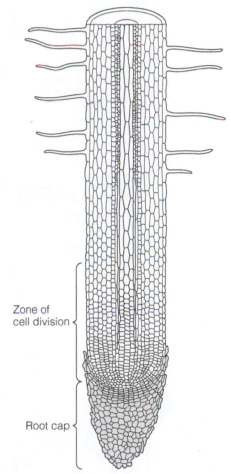

Activity B: Viewing Mitosis in Onion Root Tip Cells

Plants have regions of cell division where growth occurs. One such region is the root tip, whose growth enables the roots to elongate and reach through the soil. Since these tips are fine, threadlike tissues, it is easy to press them flat on a slide to view them under a microscope. Mitosis is concentrated in one particular region of the root tip (see Figure 8.2). After the new cells are produced, they will enlarge and mature into different cell types in the root. To produce these root tips, small white onions were placed in water. After several days of root growth, the root tips were cut off and preserved. They were then placed in a staining solution that makes their chromosomes visible.

Procedure

1. Use forceps to pick a root tip out of the vial and place it on a microscope slide.
2. Add a small drop of water to the slide.
3. Place a coverslip over the root tip.
4. Use the eraser end of a pencil to press gently on the coverslip. This should spread the tissues apart. Go easy at first—you don't want to crack the coverslip, and you can always press again if necessary.
5. Place the root tip slide on the stage of your microscope and find the pointed end of the root using scanning power.

6. Turn to low power to locate the region of cell division. If the cells do not look clear, you have not squashed the root tip enough. When it is properly squashed you will be able to see a single layer of cells.

7. Once you have located the area of cell division and are able to see cells clearly, position one cell in the center of the lens and turn to high power. You should now be able to see the chromosomes. Identify which (if any) stage of mitosis the cell is in and draw it in the appropriate box below.

8. Move the slide around to find cells in different phases of mitosis. You may need to switch back and forth between high and low power to do this.

Keep in mind that mitosis is a continuous process. Some cells will be between the "snapshot" phases.

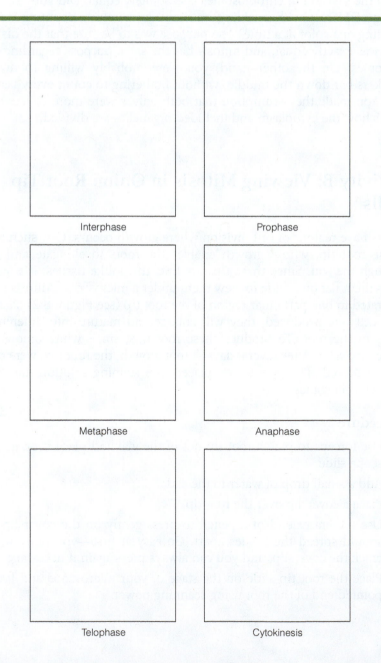

Interphase   Prophase

Metaphase   Anaphase

Telophase   Cytokinesis

EXERCISE 8.3
The Process of Meiosis

Objectives

After completing this exercise, you should be able to

1. Explain the purpose and location of meiosis in organisms.
2. Describe and recognize the phases of meiosis.
3. Explain how the process of meiosis results in genetic variation of offspring.

Mitosis produces an exact copy of a cell and is an essential process for growth and repair in multicellular organisms. But multicellular organisms also need a special type of cell division to produce gametes (eggs and sperm), and that process is called meiosis. Unlike mitosis, which occurs throughout the body, meiosis occurs only in specialized reproductive tissues.

One outcome of meiosis is reduction of chromosome number. Look again at Figure 8.1, the human karyotype, which shows the homologous pairs of a diploid cell. Gametes made from this cell will be *haploid* with 23 chromosomes. How should 23 chromosomes be selected to go into each gamete?

Enormous genetic variation of gametes is another outcome of meiosis. On the karyotype in Figure 8.1, randomly label each chromosome as maternal (M) or paternal (P). Then randomly choose one member of each pair to put in a gamete (or flip a coin to choose) and write your selections (M or P) below for each chromosome.

— —
  1 2 3 4 5 6 7 8 9 10 11 12 13 14 15 16 17 18 19 20 21 22 sex

Again, each chromosome is a DNA molecule that encodes information and each segment of information is called a gene. So distributed on the 46 chromosomes in the karyotype in Figure 8.1 are the 30,000 to 40,000 genes that make up the human genome, the set of instructions that produces a human being. In the next lab topic you will consider the distribution of genes during meiosis, but for now we are going to examine how the chromosomes themselves are sorted into gametes.

Activity A: Pop Bead Simulation of Meiosis

To see how the process of meiosis results in haploid gametes that differ from each other genetically, you will take a cell through the steps of meiosis as you did for mitosis. The mechanisms of mitosis, such as the centrosomes and spindle fibers, are also used in meiosis, and the same names—prophase, metaphase, anaphase, telophase—are given to the "snapshot" phases. But meiosis will result in four cells rather than two, so two divisions take place. The two divisions are designated meiosis I and meiosis II. Also, while many different types of cells can undergo mitosis, only specific cells can undergo meiosis. In animals, meiosis is carried out in the cells that produce gametes. In plants, meiosis produces spores (Lab Topic 12).

Where in the human body does meiosis occur?

Procedure

Begin with the same three homologous pairs of chromosomes that you used for mitosis.

Interphase
Before meiosis begins, the chromosomes must replicate. Make an exact copy of each of your chromosomes, then draw the interphase nucleus in the cell below.

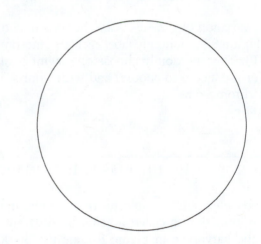

Prophase I
As in mitosis, the chromosomes become visible and the nuclear envelope disintegrates during this phase. But something very different happens, too. Homologous pairs associate closely with each other, even to the point of entwining their "arms." Each pair of pairs is called a **tetrad.** Group your chromosomes into tetrads.

While they are in close contact, the chromatids of homologous chromosomes can exchange pieces of DNA. The size of these exchanged segments varies. You can simulate this process, called crossing over, by popping a few beads off of one chromatid and exchanging them for the same number of beads on a chromatid from its homologue.

What is the result of crossing over in a homologous pair?

Sketch prophase I below.

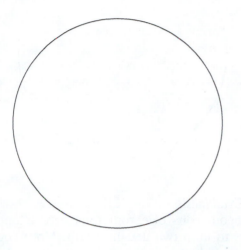

Metaphase I

Recall that metaphase of mitosis is when the chromosomes are attached to the spindle apparatus and aligned along the center of the cell. Metaphase I in meiosis is similar, but each tetrad stays together. Line up your tetrads and sketch them below.

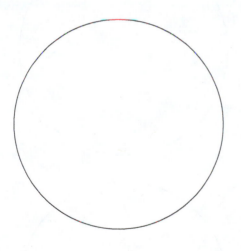

Anaphase I

In anaphase I, homologous pairs are separated from each other and move to opposite poles, but sister chromatids remain attached. How does this differ from anaphase of mitosis?

Move your chromosomes to simulate anaphase I and sketch it below.

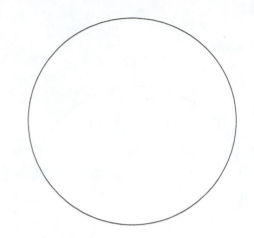

Telophase I and Cytokinesis
Once the homologous pairs of chromosomes are at opposite poles, the cytoplasm divides to form two daughter cells. Move your chromosomes into the two new daughter cells that will be formed when cytokinesis is completed. Make a sketch below.

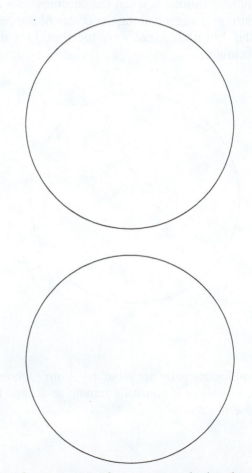

List by color the chromosomes that are in each daughter cell you made. For example, if your chromosomes are red and yellow, your daughter cells might be red, red, yellow and yellow, yellow, red. (For chromosomes that underwent crossing over, list the predominant color.)

Check with other students to see whether their daughter cells are the same as yours. If there are different combinations, can you explain why? (Hint: Review the metaphase I sketch of someone whose daughter cells turned out differently than yours.)

Crossing over during prophase I is one source of genetic variation that results from meiosis. The random alignment of tetrads at metaphase I is another source. With only three pairs of chromosomes, there are eight different combinations of chromosomes in the gametes (two for each possible alignment). Imagine how many possibilities exist with 23 pairs of chromosomes!

Each cell that results from meiosis I has one member of each homologous pair, which is the desired outcome for gametes. Why isn't this the end of meiosis?

Prophase II
During prophase II, a spindle apparatus forms again.

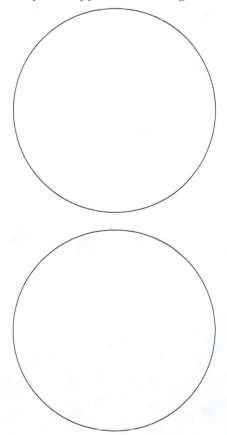

Metaphase II
With each chromatid attached to spindle fibers, the chromosomes line up in the center of the cell.

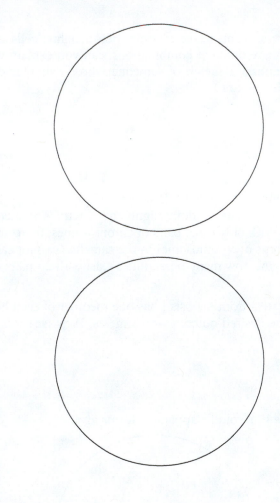

Anaphase II
During this anaphase, the sister chromatids move to opposite poles.

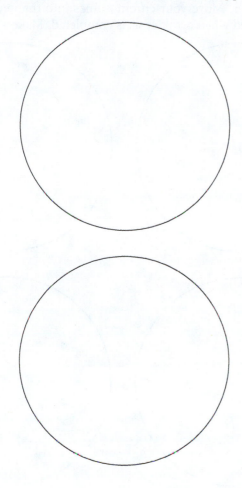

Telophase II and Cytokinesis
The division of chromatids is completed, and the cytoplasm divides to form new daughter cells. Move your chromosomes into the new daughter cells that will be formed when cytokinesis is completed. Make a sketch below.

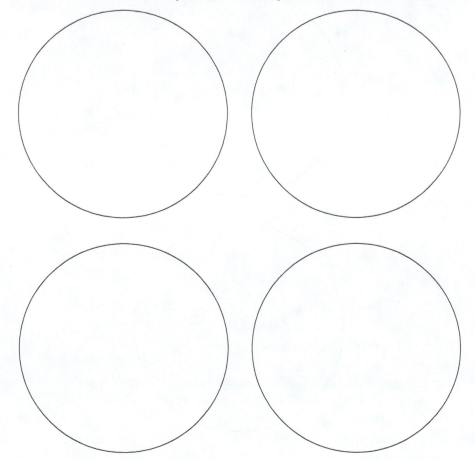

How many chromosomes does each cell have?

How many homologous pairs does each cell have?

Are these cells haploid or diploid?

List by color the chromosomes that are in each daughter cell you made.

Check with other students to see whether their gametes turned out the same as yours. Make a complete list of all possible chromosome combinations in the gametes.

Activity B: Viewing Meiosis in Organisms

You will view professionally prepared slides to see what actual cells look like during the phases of meiosis. If you were to use animal tissue for this exercise, what parts of the animal would be used to show meiosis?

If you were to use plant material, which parts would you look at to see meiosis? (Hint: Where does sexual reproduction take place?)

Procedure

1. Your instructor will supply you with prepared microscope slides to examine. Begin by using scanning power to locate the specimen. Switch to low power to identify an area that includes cells undergoing meiosis.

2. Using high power, sketch the positions of the chromosomes during different stages of meiosis in the spaces below. Label your drawing with the name of the stage, as best you can identify it.

Why isn't it as easy to identify the stages of meiosis in cells as it was to identify the stages of mitosis?

Questions for Review

1. What purpose does mitosis serve in development and growth? (Hint: A zygote is the first cell formed after conception, but by adulthood a human body has many trillions of cells.)

2. After an adult has all his trillions of cells, does he still need the process of mitosis? Explain.

3. Which phase(s) of mitosis could the karyotype in Figure 8.1 have been made from? Explain your reasoning.

4. Besides the root tips, where else in plants would you expect mitosis to take place?

5. Meiosis is from a Greek root meaning "to diminish or lessen." Why is this an appropriate term?

6. Summarize the sources of genetic variation in meiosis.

7. Which phase of meiosis is responsible for the genetic variation of gametes demonstrated in the introduction to Exercise 8.2?

8. Most people who have Down syndrome have an extra copy of chromosome 21 (trisomy 21), which is a result of an egg or sperm that carried an extra copy of chromosome 21. How could a gamete with an extra chromosome come about?

9. You have homologous pairs of chromosomes because you got one member of each pair from each parent. Your parents got their chromosomes from your grandparents, and so on. But are the chromosomes that are passed along through the generations exactly the same? From example, is your maternal chromosome #1 exactly the same one that your mother received from one of her parents? Explain.

Mendelian Genetics

Introduction

The study of genetics involves learning how traits are passed from one generation to the next. Humans have been interested in genetics since the beginning of agriculture. The modern science of genetics began in the 1860s with investigations by Gregor Mendel. His experiments with pea plants provided the foundation for understanding mechanisms of heredity. In this lab topic you will learn how the pattern of inheritance of some traits follows the principles discovered by Mendel. You will also perform a cross using Wisconsin Fast Plants™, which are members of the mustard family, and interpret the results according to Mendelian principles.

 Bring a calculator for this lab topic.

Before coming to this lab, refer to your textbook to write a definition for each of the following terms:

Gene:

Locus:

Allele:

Dominant allele:

Recessive allele:

Genotype:

Phenotype:

Homologous chromosomes:

Diploid:

Haploid:

Heterozygote:

Homozygote:

Outline

Exercise 9.1: Mendel's Principles

 Activity A: The Chromosomal Basis of Segregation

 Activity B: Predicting the Outcome of a Monohybrid Cross

 Activity C: The Chromosomal Basis of Independent Assortment

 Activity D: Predicting the Outcome of a Dihybrid Cross

Exercise 9.2: A Dihybrid Cross Using Wisconsin Fast Plants

 Activity A: Planting F_1 Seeds

 Activity B: First Observations

 Activity C: Pollination

 Activity D: Germinating the F_2 Seeds

 Activity E: Collecting and Interpreting Data from the F_2 Generation

E X E R C I S E   9 . 1
Mendel's Principles

Objectives

After completing this exercise, you should be able to

1. Define and explain Mendel's principle of segregation.
2. Define and explain Mendel's principle of independent assortment.
3. Follow one pair of alleles through meiosis and determine the resultant gametes from a diploid cell with two homologous pairs of chromosomes.
4. Follow two unlinked pairs of alleles through meiosis and determine the gametes (or spores) that result from a diploid cell with two homologous pairs of chromosomes.
5. Solve problems involving monohybrid and dihybrid crosses.

Activity A: The Chromosomal Basis of Segregation

Diploid cells contain two sets of **homologous chromosomes.** One set, or one member of each pair, comes from each parent. Each pair of homologous chromosomes carries **genes** that govern the same traits. For example, in pea plants, flower color is determined by a single gene F, which can have two different forms, F or f, called **alleles.** A gene is a segment of DNA, and alleles have different sequences of nucleotides in that segment. Every cell in the diploid plant has two copies of the gene, one on each member of a homologous pair of chromosomes. These two versions of the same gene may be alike (**homozygous**) or different (**heterozygous**). The genetic makeup of the cell is known as its **genotype.** In this example, a cell with two F alleles would have the genotype FF. A cell with two f alleles would have the genotype ff. If the cell has one F and one f, its genotype is Ff. The genotypes FF and ff are homozygous, and the genotype Ff is heterozygous. All of the cells of the plant should have the same genetic composition.

In pea plants, the F allele is **dominant** over f: When one F is present it masks the f allele. The f allele is called **recessive.** A capital letter is used to denote the dominant allele. A pea plant having either the FF or Ff genotype has the **phenotype,** or outward appearance, of purple flowers. The phenotype that results from the ff genotype is white flowers.

Genetic traits are passed from one generation to the next by reproduction. When animal cells undergo meiosis (Lab Topic 8) they produce **gametes,** which are haploid. When plant cells undergo meiosis they produce **spores,** which then become the plants that produce gametes. This difference between plant and animal reproduction is explained in detail in Lab Topic 12 (Plant Diversity).

In the next example we will follow a gene R that codes for the tongue-rolling ability in humans. The allele R, which gives a person this ability, is dominant over r.

What is the phenotype of an individual whose genotype is RR?

Figure 9.1.
Tongue-roller phenotype. This person can roll her tongue into a U-shape.

What is the phenotype of an individual whose genotype is *Rr?*

What is the phenotype of an individual whose genotype is *rr?*

What are your phenotype and genotype?

After meiosis, each resultant gamete will have only one member of each homologous pair of chromosomes. Therefore each gamete will have only one allele for tongue-rolling.

The distribution of alleles during the formation of gametes was one of the principles described by Gregor Mendel. It is called the **principle of segregation:** The two alleles of a gene segregate, or separate, from each other so that each one ends up in a different gamete. Mendel, however, did not know about the existence of chromosomes or meiosis. Not until decades after his death was the chromosomal basis of Mendel's law discovered.

If a person's genotype is *RR*, what are the genotypes of the resulting gametes?

If the person's genotype is *rr,* what are the genotypes of the resulting gametes?

If the person's genotype is *Rr,* what are the genotypes of the resulting gametes?

Activity B: Predicting the Outcome of a Monohybrid Cross

When the genotypes of the parents are known, we may determine what gametes the parents can make and in what proportion the gametes will occur. This information allows us to predict the genotypes and phenotypes of the offspring. The prediction is simply a matter of listing all of the possible combinations of gametes. In this section you will be doing **monohybrid** crosses: Only one trait is followed.

By convention, the parental generation is called P. The first generation of offspring is called F_1. F stands for filial, which refers to a son or daughter, so F_1 is the first filial generation. If members of the F_1 generation are crossed, their offspring are called the F_2 generation, and so on.

Predict the results of the following cross (using R to denote tongue-rolling ability):

P generation: $RR \times RR$

What genotype(s) will be found in the F_1 generation?

What phenotype(s) will be found in the F_1 generation?

Explain why you made these predictions.

Predict the results of the following cross:

P generation: $RR \times rr$

What genotype(s) will be found in the F_1 generation?

What phenotype(s) will be found in the F_1 generation?

Explain why you made these predictions.

In the examples given so far, each parent has only been able to produce one type of gamete, so the outcomes of the crosses are fairly simple. The **Punnett square** was devised to keep track of all possible combinations of genotypes when more than one type of gamete can be produced. Fill out the Punnett square in Figure 9.2 for the F_2 generation by crossing offspring of the previous cross ($Rr \times Rr$).

What are the possible genotypes in the F_2 generation?

What is the phenotype of each genotype in the F_2 generation?

What is the phenotypic ratio for this cross?

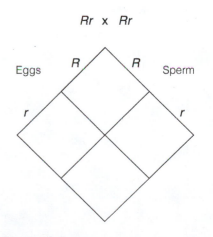

Figure 9.2.
Punnet square for monohybrid cross.

Activity C: The Chromosomal Basis of Independent Assortment

Genes that are located on the same chromosome are **linked** with each other. If genes are located on separate, nonhomologous chromosomes, they are not linked, or unlinked. Unlinked genes separate independently during meiosis. For example, consider the allelic pair *R* and *r* and a second allelic pair *A* and *a*. If the *R* gene and the *A* gene are not linked, their alleles can be found in any combination in the gametes. That is, the *R* allele can be in the same gamete as either *A* or *a*. This is Mendel's **principle of independent assortment.** The word assortment in this case refers to the distribution, or sorting, of alleles into gametes.

In Figure 9.3, draw a cell that represents the following conditions: diploid; two homologous pairs of chromosomes; two unlinked genes called *R* and *A*; cell is heterozygous for both genes.

What is the genotype of this cell?

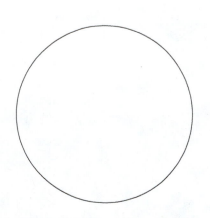

Figure 9.3.
Diploid cell carrying two unlinked genes.

R is the gene for tongue-rolling, as used in the previous examples. *A* determines arch characteristics. A person who has the dominant allele has normal arches. An individual who is homozygous recessive has flat feet.

What is the phenotype of the individual represented by the cell in Figure 9.3?

Recall that when this cell undergoes meiosis, each gamete receives one member of each homologous pair. List the possible combinations of alleles that will be found in the gametes.

In what proportion would you expect these gametes to occur?

Activity D: Predicting the Outcome of a Dihybrid Cross

The outcomes of **dihybrid** crosses (following two traits) can be determined just as they are for monohybrid crosses. List all possible gametes each parent can produce and make all possible combinations of those gametes. When determining what the gametes will be, remember that each gamete must have one member of each homologous pair of chromosomes. If you are considering an *A* gene and an *R* gene, for example, each gamete must have one allele for the *A* gene (either *A* or *a*) and one allele for the *R* gene (either *R* or *r*). Try the following examples.

Genotype: *RRAA*   Gametes: _____

Genotype: *RrAA*   Gametes: _____

Genotype: *rrAa*   Gametes: _____

Genotype: *RrAa*   Gametes: _____

To predict the outcome of a cross, construct a Punnett square to accommodate the gametes from each parent. For example, cross two heterozygous parents using the Punnett square in Figure 9.4.

RrAa x *RrAa*

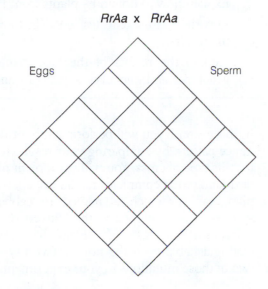

Eggs   Sperm

Figure 9.4.
Punnett square for dihybrid cross.

List the possible genotypes of the offspring:

List the possible phenotypes of the offspring and the expected ratios in which they would occur.

A couple with the genotypes *RrAa* and *RrAa* have 16 children. Ten of them have normal arches and can roll their tongues; the other six have normal arches and can't roll their tongues. Why doesn't this family match the expected ratio? (Hint: If the probability of having a male child is $\frac{1}{2}$, why can one family have seven daughters and no sons?)

EXERCISE 9.2
A Dihybrid Cross Using
Wisconsin Fast Plants

Objectives

After completing this exercise, you should be able to

1. Identify the wild type and anthocyaninless (green stem) and yellow-green mutant forms of Fast Plants.

2. Explain how to obtain F_2 plants from F_1 seeds.

3. Perform a chi-square test on data from a genetic cross and interpret the results.

4. Explain the results of the Fast Plants experiment in terms of the genotypes and phenotypes of the F_1 and F_2 generations.

In this exercise you will perform a dihybrid cross and determine the inheritance pattern by comparing your results to the expected phenotypic ratio for this type of cross. You will use a plant of the genus *Brassica* in the mustard family that completes its life cycle faster than most other plants. This plant variety, which is related to cabbage, cauliflower, and brussels sprouts, was developed at the University of Wisconsin and is known as Wisconsin Fast Plants. Fast Plants have several mutations that are clearly distinguishable from the normal (wild type) for the variety. You will use two of those mutations in your experiment.

The stems of wild-type seedlings have a red or purplish tinge that is caused by anthocyanin pigments. If you performed Lab Topic 2, pH and Buffers, you may recall using the anthocyanin pigments of red cabbage as a pH indicator. The mutant form does not produce any anthocyanin, so it has a green stem.

If the mutation is a simple recessive gene, what are the possible genotypes of a plant that produces anthocyanin?

What is the genotype of a plant that does not produce anthocyanin?

Like most plants, the leaves of Fast Plants produce plenty of chlorophyll, which makes them appear green. A mutant form has yellowish-green leaves.

If the mutation is a simple recessive gene, what are the possible genotypes of a plant that has green leaves?

What is the genotype of a plant that has yellowish-green leaves?

What is the genotype of a plant that is heterozygous for both genes?

What is the phenotype of a heterozygote?

If both mutations are unlinked, simple recessive alleles and two heterozygous parents are crossed, what ratio of phenotypes is expected in the offspring? See Figure 9.4 if you need to review the expected outcome of a dihybrid cross.

In this exercise you will perform a dihybrid cross between two heterozygotes and compare the phenotypic ratio of the offspring to the expected ratio. Fast Plants are only fast relative to other plants. In fact, the experiment takes about 7 weeks to complete. You will perform different stages of the experiment during several labs prior to doing Lab Topic 9, when you will collect data.

You will perform the following steps:

- Plant the seeds, which represent the F_1 generation.
- Record the phenotypes of the seedlings.
- When the plants flower, cross-pollinate them.
- Wait for the seeds to mature and harvest them.
- Plant the seeds, which represent the F_2 generation.
- Record the phenotypes of the F_2 seedlings.

During this time, of course, the plants must be watered to remain healthy. You will be responsible for ensuring that your plants survive to produce data.

Activity A: Planting F_1 Seeds

7 Weeks Before Activity E

Your lab team will be responsible for one quad. A quad is a very small styrofoam pot with four compartments (Figure 9.5a). In order for Fast Plants to complete their life cycle rapidly, they require special growing conditions. Fertilizer pellets will be added to the soil, and the plants will need to receive constant light.

Procedure

1. Push one wick through the hole in the bottom of each compartment so that the tip sticks through about 1 cm (Figure 9.5a).
2. Add potting soil so that each compartment is about half full.
3. Place three fertilizer pellets in each compartment.

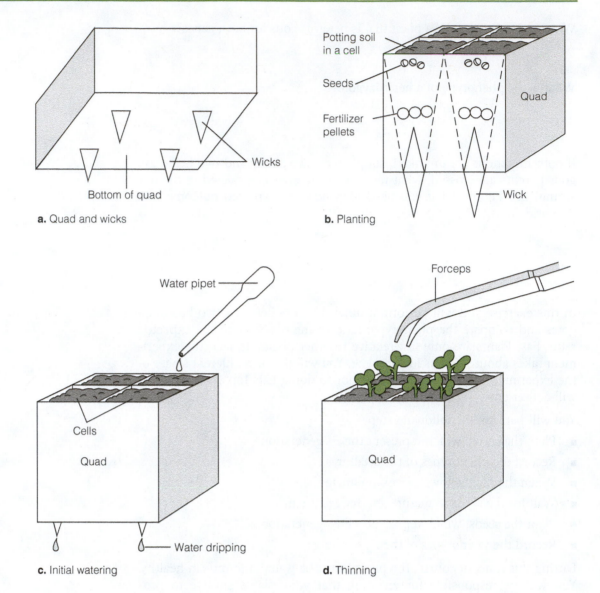

a. Quad and wicks

b. Planting

c. Initial watering

d. Thinning

Figure 9.5.
Planting and thinning Fast Plants.

4. Add potting soil so that each compartment is loosely filled.

5. Use a fingertip to make a slight depression in the soil of each compartment.

6. Put three seeds in each depression.

7. Sprinkle a little more potting soil in each compartment so that the seeds are covered.

8. Using a waterproof pen, write the date and your team's initials on a pot label and insert it in one of the compartments of your quad.

9. Use a pipet to water each compartment gently until you see water dripping from the wicks (Figure 9.5c).

10. Place your quad under the lights on the water mat along with those of other lab teams. The water mat draws water from the reservoir beneath it. In turn, the wicks will draw water into each quad.

Caring for Plants

First 3 Days After Planting

Plants should be watered from above with a pipet for the first three days after planting (Figure 9.5c). This may be handled by your instructor. It is critical to keep the soil moist while the seeds are germinating.

Daily Thereafter

Check the soil moisture in your quad daily. Your instructor may do this for you. If it is dry, pick up the quad and add water with a pipet until water drips from the wick. If the mat beneath your quad is dry, ask your instructor to resoak it.

Even if your plants have wilted, don't give up. Water the plants from above with a pipet. Then place the quad in about 1 cm of water in a sink or dish tub and leave it until the plants recover. However, if the plants are already brown and crispy, it is too late to save them.

If the plants will not stand upright as they grow taller, you may support them with a small wooden stake.

Activity B: First Observations
6 Weeks Before Activity E

When the seedlings are approximately 1 week old, compare them to the wild-type plants grown by your instructor to be sure that they express the correct phenotype. Destroy any plants that lack anthocyanin in the stems or have yellow-green leaves. That is, destroy any plants that express the mutant phenotype. Why should these plants not be used for your experiment?

Thin the plants so that there is only one per compartment in your quad (Figure 9.5d). Leave the most vigorous-looking plant to survive.

Preparation for Pollination

Each group should prepare a bee stick to be used to transfer pollen from one flower to another. As you will learn in Lab Topic 18, Flowers, Fruits, and Seeds, bees are natural pollinators, so their bodies contain small hairs that pick up pollen easily. Alternatively, a soft paintbrush can be used for pollination.

 Do not perform this procedure near the plants. The glue may contain a chemical that will prevent pollen germination.

Procedure

1. Hold a dead bee by its wings. Remove its head, legs, and abdomen. The part remaining is the thorax. The thorax will have a small hole at each end where the other body parts were removed.
2. Put a small drop of glue on the tip of a toothpick.
3. Insert the tip of the toothpick into the hole at one end of the thorax.
4. Stick the other end of the toothpick into a piece of styrofoam and allow the glue to dry.

Activity C: Pollination

Approximately 5 Weeks Before Activity E

Flowers will begin to open about 2 weeks after planting. Pollinate three times, 2 days apart. You may exchange pollen with the plants in your quad or with plants in other quads.

✳ **Do not cross-pollinate with the demonstration plants grown by the instructor.**

Procedure

1. Rub the bee thorax gently over the anthers (male part) of a flower to obtain pollen.
2. Transfer the pollen to a new plant by brushing the bee on the stigma (female part) of a flower on another plant.
3. Repeat 2 days and 4 days later.
4. After you have pollinated three times, remove all new flower buds that appear so that only the flowers that you have pollinated will produce seeds. Check plants two or three times a week for this purpose. This may be handled by your instructor.

Harvesting Seeds

It takes approximately 3 weeks for the seed pods to mature. During that time, you or your instructor should continue to ensure that your plants have enough water and continue to remove new flower buds and shoots.

Two weeks before the data collection lab (approximately 3 weeks after pollination), remove the plants from the watering system so the seed pods will dry out.

Activity D: Germinating the F_2 Seeds

1 Week Before Activity E

The seeds that you harvest are the F_2 generation. In order to determine the phenotypes of these plants, you will germinate them on filter paper. There is no need to grow the plants in soil as you did with the F_1 generation because you will be able to determine the phenotypes from week-old seedlings.

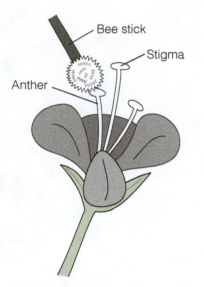

Figure 9.6.
Reproductive structures of Fast Plants.

Procedure

1. Remove the dry seed pods from each plant.
2. Over a collecting pan, roll each pod between your thumb and forefinger to remove the seeds.
3. Use a pencil to write the date and your lab team's initials on one end of a filter paper disk.
4. Place the filter paper in the top of a petri dish, which is larger than the bottom.
5. Use a wash bottle to wet the filter paper thoroughly.
6. Place approximately 25 seeds in rows across the top two-thirds of the dish.
7. Fit the bottom of the petri dish into the top and stand it upright in a reservoir (Figure 9.7). There should be about 2 cm of water at the bottom of the reservoir.
8. Empty the contents of each quad into the container provided.
9. Return the quad and the label to your instructor.

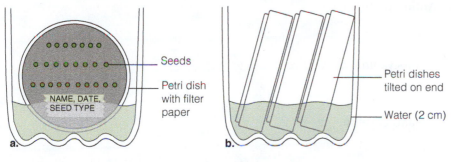

Figure 9.7.
Germination of Fast Plant seeds.

Activity E: Collecting and Interpreting Data from the F_2 Generation

Data Collection Lab

Procedure for Collecting Data

1. Count the number of seedlings that show each of the phenotypes listed in Table 9.1. Score a seedling as having anthocyanins if the stem has even a slight reddish or purple tinge. Ask your instructor to show you examples of each phenotype if you are in doubt. Using forceps, remove each seedling from the filter paper as it is counted so you won't count it again.
2. Record your data in Table 9.1 on the next page.

As discussed in an earlier section of the lab, when there is a small number of observations the data may not display the expected phenotypic ratio. For example, if the parents are both heterozygous for tongue-rolling, they may have two children who are tongue-rollers and two who are not instead of the 3:1 ratio expected of a monohybrid cross. What is the

Table 9.1
Data from Dihybrid Cross Between *AaGg* Parents

Phenotype	Number Observed (your data)
Anthocyanins (purple stem), green leaves	
Anthocyanins (purple stem), yellow-green leaves	
No anthocyanins (green stem), green leaves	
No anthocyanins (green stem), yellow-green leaves	

expected phenotypic ratio in the dihybrid cross you performed with the Fast Plants? (Hint: Review the Punnett Square in Figure 9.4.)

Anthocyanin, green:

Anthocyanin, yellow-green:

No anthocyanin, green:

No anthocyanin, yellow-green:

Chi-square is a statistical test that determines how well the actual data fit the expected ratio. The test is performed by calculating a value called chi-square. A probability table is then used to determine whether this value fits the expected ratio. Small deviations from the expected ratio can happen due to chance. The chi-square test tells us whether the difference between the observed data and the expected data can be attributed to chance. If it can, then the data *do* fit the expected ratio for the F_2 generation of a dihybrid cross; the deviation is not significant. If the deviation is not due to chance, then it must be explained some other way. For example, a large deviation could mean that the inheritance does not follow a simple dominant-recessive pattern.

The example in Table 9.2 shows how to calculate chi-square. The data recorded from the experiment, the number of each phenotype observed, is recorded in the first column. Then the expected number of each phentoype is calculated, based on the total number of seedlings that were counted (177 in the example). The total number is multiplied by the expected ratio for each phenotype. Then the deviation of the observed number from the expected number in each category is calculated. Finally the deviations from all four categories are added. The result is the chi-square value.

Table 9.2
Example Calculation of Chi-Square

Phenotype	Number Observed	Number Expected	Deviation = $\dfrac{(\text{Observed} - \text{Expected})^2}{\text{Expected}}$
Anthocyanins, green	102	$177 \times {}^{9}/_{16} = 100$	0.04
Anthocyanins, yellow-green	40	$177 \times {}^{3}/_{16} = 33$	1.48
No anthocyanins, green	27	$177 \times {}^{3}/_{16} = 33$	1.09
No anthocyanins, yellow-green	8	$177 \times {}^{1}/_{16} = 11$	0.82
Total seedlings observed = 177.		Sum of all deviations = chi-square = 3.43.	

The chi-square value is then compared to Table 9.3 to determine whether the observed deviation from the expected ratio is likely to have occurred by chance. Table 9.3 is an abbreviated version of a chi-square table that includes only the row of values that pertain to your data. If you want to perform a chi-square test for some other purpose, you should obtain a complete table from a statistics book.

If the probability is greater than 0.05, we can say that the differences between the observed data and the expected ratio are due to chance. In the example, the chi-square value of 3.43 corresponds to a probability of between 0.50 and 0.30. This tells us that the observed ratio is not significantly different from the expected ratio. The data in this example would support the hypothesis that the anthocyaninless and yellow-green phenotypes are caused by a recessive allele.

Procedure for Interpreting Data

1. Record the class data in the first column of Table 9.4.
2. Add up the total number of seedlings observed for the experiment.
3. Calculate the expected number for each phenotype. To do this, multiply the total number by ${}^{9}/_{16}$, ${}^{3}/_{16}$, or ${}^{1}/_{16}$, as shown in Table 9.2. This calculation determines the share of the total that is expected for each phenotype. Remember, the expected ratio is derived from the Punnett square for a dihybrid cross (Figure 9.4).

Table 9.3
Chi-Square Table for Dihybrid Cross

	Probability That Deviation Occurred by Chance										
	0.95	0.90	0.80	0.70	0.50	0.30	0.20	0.10	0.05	0.01	0.001
Chi-Square =	0.35	0.58	1.01	1.42	2.37	3.66	4.64	6.25	7.82	11.34	16.27
	Deviation from expected ratio is explained by chance.								Deviation is too great to be due to chance.		

Source: Fisher, R. A., and F. Yates. *Statistical Tables for Biological, Agricultural and Medical Research,* 6th ed., Table IV. Edinburgh: Oliver and Boyd, Ltd.

Table 9.4
Chi-Square Test of Class Data

Phenotype	Number Observed (class data)	Number Expected	Deviation = $\frac{(Observed - Expected)^2}{Expected}$
Anthocyanin, green		$\underline{\hspace{2cm}} \times {}^9/_{16} = \underline{\hspace{2cm}}$ (total #) (# exp.)	
Anthocyanin, yellow-green		$\underline{\hspace{2cm}} \times {}^3/_{16} = \underline{\hspace{2cm}}$ (total #) (# exp.)	
No anthocyanin, green		$\underline{\hspace{2cm}} \times {}^3/_{16} = \underline{\hspace{2cm}}$ (total #) (# exp.)	
No anthocyanin, yellow-green		$\underline{\hspace{2cm}} \times {}^1/_{16} = \underline{\hspace{2cm}}$ (total #) (# exp.)	
Total seedlings observed (class) = _____.		Sum of all deviations = chi-square = _____.	

4. Calculate the deviations of the observed values from the expected values using the formula shown in the last column of Table 9.4.

5. Find the sum of the deviations. This number is the chi-square value.

6. Compare the chi-square value to the second row of Table 9.3 to determine the probability that the observed deviations can be explained by chance.

What is the probability?

Do your results provide evidence that the anthocyaninless and yellow-green mutants are recessive alleles? Explain your answer.

Questions for Review

To practice doing monohybrid and dihybrid crosses, work the following problems. Each of the human traits used in the problems is determined by a single gene locus that has two different alleles, one dominant and one recessive. Show all of your work for the problems.

Dominant traits	Recessive traits
Freckles (F)	No freckles (f)
Astigmatism (A)	Normal vision (a)
Ability to roll tongue (R)	Cannot roll tongue (r)
Normal arches (A)	Flat feet (a)
Widow's peak (W)	Straight hairline (w)

1. A man who has normal vision marries a woman who is heterozygous for astigmatism. What are the possible genotypes and phenotypes that their children can have?

2. Two people who have normal arches produced a child who has flat feet. What is the genotype of the child?

    What is the genotype of the father?

    What is the genotype of the mother?

    Explain how you arrived at your answers.

3. A woman has a widow's peak, but she does not know her genotype. She marries a man who has a straight hairline and they have 13 children. Nine have widow's peaks and four have straight hairlines. What are the genotypes of the parents? What are the genotypes of the children? Explain how you arrived at your answers.

4. A couple who both have the ability to roll their tongues have a son who is also a tongue-roller. The son is very curious as to whether he is homozygous or heterozygous for the tongue-rolling trait. How would he go about finding out?

5. A man who has no freckles and flat feet marries a woman who is homozygous dominant for both traits. What is the man's genotype?

   What genotype(s) will his gametes have?

   What is the woman's genotype?

   What genotype(s) will her gametes have?

   What genotype(s) will their children have?

   What will their children's phenotype(s) be?

6. A woman who is heterozygous for astigmatism and cannot roll her tongue marries a man who has normal vision and is homozygous dominant for tongue-rolling. What are the possible genotypes and phenotypes of their children for these traits?

7. A man has freckles and can't roll his tongue. His wife has no freckles and can roll her tongue. All of their many children have freckles. About half can roll their tongues, while the other half can't. What are the probable genotypes of the parents, and what gametes can each parent produce?

8. A woman who has normal arches and a straight hairline has children with a man who has flat feet and a widow's peak. One of their children has a straight hairline and normal arches, one of their children has a straight hairline and flat feet, one has a widow's peak and normal arches, and one has a widow's peak and flat feet. What are the genotypes of the parents?

9. The F_1 seeds for the Fast Plants experiment were heterozygous for both traits (anthocyanin and leaf pigment). How were these heterozygote seeds produced?

Forensic Application of Molecular Genetics

✳ You will need a calculator for this lab!

Introduction

You already know that DNA is the genetic material and that its sequence of nucleotides holds the "recipe" for making proteins. But it turns out that only about 3% of our genes are actually the instructions to make functional proteins. The rest of the nucleotide sequences are nonsensical or irrelevant. Collectively, they are known as "junk" DNA. Like old magazines or newspapers that have accumulated over decades in someone's attic, there is no use for junk DNA in our bodies. But scientists have found a way to turn trash into treasure, based on the fact that each of us has our own unique collection of junk. Although we vary genetically in many of the functional genes, we have even more variability in certain types of these nonsense sequences. There are thousands of sites in our chromosomes where they have been inserted. Although numerous people may share the same bit of junk at a single site, we are extremely unlikely to share the same ones at each of several sites. Statistical comparisons are used to determine how likely it is that two samples of DNA came from the same source. Thus, some of the most effective techniques for DNA typing—otherwise known as DNA fingerprinting or DNA profiling—are based on our unique patterns of junk DNA.

In this laboratory, you will learn about one of the methods of DNA typing that is currently widely used. You will then learn how to compare the results of DNA analysis to statistics for a population group. To complete the laboratory, you will apply your knowledge of DNA profiling to solving a crime.

Outline

Exercise 10.1: DNA Profiling Using STR Analysis

    Activity A: The STR Technique

    Activity B: Application of STR Analysis to Paternity Testing

    Activity C: Using Population Genetics to Match DNA

Exercise 10.2: Application of DNA Profiling

E X E R C I S E   1 0 . 1
DNA Profiling Using STR Analysis

Objectives

After completing this exercise, you should be able to

1. Explain what STRs are and why they are useful for DNA profiling.
2. Describe the role of restriction enzymes in STR analysis.
3. Describe the role of electrophoresis in STR analysis.
4. Apply the laws of Mendelian inheritance to STR alleles to determine parentage.
5. Use allele frequencies to calculate the probability of an individual having a particular genotype.

Activity A: The STR Technique

One type of junk DNA that has proven useful in DNA profiling is called STR (simple tandem repeat). A STR (pronounced "star") is a locus or place in the DNA where a short segment of nucleotides—for example, ACG—is repeated numerous times. We all share the same sequence of repeated nucleotides (ACG, in this case) but the *number* of repetitions varies. For example, in one allele there might be seven repetitions: ACGACGACG ACGACGACGACG. In another allele there might be 25 repetitions—ACG ACGACGACGACGACGACGACGACGACGACGACGACG ACGACGACGACGACGACGACGACG—while a third allele might have 33 repetitions. More than 8,000 STR loci have been identified in the human genome.

In contrast to the simple dominant/recessive loci that you are most familiar with, which have only two possible alleles (for example, *A* and *a*), STR loci have *many* different alleles. As a result, the frequency of any particular allele within a population may be quite low. For example, at one locus called D3S1358 that is used in DNA profiling, the Caucasian population has 11 possible alleles whose frequencies range from 0 to 0.258. To put it another way that will be useful later in the lab, the probability that a Caucasian has a "15" allele (15 repetitions) at D3S1358 is 0.258.

Recall from Lab Topic 9 (Mendelian Genetics) that each person has the same locus on two homologous chromosomes. Just as an individual could be homozygous or heterozygous at a locus such as the one for tongue-rolling, an individual can be homozygous or heterozygous for STR loci. An individual who has two identical alleles—for example, both chromosomes have 7 repeats—would be homozygous. An individual who has seven repeats on one chromosome and 25 repeats on the homologous chromosome would be heterozygous.

The methods that scientists use to analyze an individual's alleles involve isolating the repeating segments and sorting them by size. Although in actual practice the techniques are complicated to perform, it is fairly easy to visualize the results with paper models. For the sake of simplicity, we will use single strands of DNA instead of the actual double-stranded molecule.

Procedure

Make simulated segments of DNA that contain STRs.

1. Your instructor will give you four nucleotide sequences. Cut them into single strips. These paper strips represent small portions of chromosome pairs from two individuals. Keep in mind that these segments come from much longer strands of DNA.

Isolate the STRs.
Restriction enzymes are molecules that chemically cut the DNA segments into smaller pieces by breaking bonds. There are many different restriction enyzmes. Each one "recognizes" a different base sequence in the DNA where it makes its cut.

2. You will "cut" the DNA strands with two different restriction enzymes. One enzyme recognizes "ACTATG" and cuts the DNA between the T and the G. Find that sequence in your DNA strands and cut each strip of paper there.

3. The other restriction enzyme recognizes the sequence CACTAT and cuts after the first C and before the A. Find that sequence in your DNA strands and cut each strip of paper there.

4. You have now isolated the STRs for these chromosomes. Keep only the STRs and discard the other DNA.

What is the STR sequence?

Sort the STRs by size.
Scientists sort the pieces of DNA with a technique called electrophoresis that uses a slab of gel-type material. DNA samples are lined up in lanes on the gel like a row of swimmers on the blocks before a race. The gel is then subjected to an electrical field, which causes the DNA samples to travel in a straight line from their starting points. The gel resists the molecules as they move through it, so shorter DNA segments travel farther than longer ones. After the samples have been "run," techniques to visualize specific bands of DNA are used. An example is shown in Figure 10.1. Notice that a "standards" lane is included on the gel. Pieces of DNA of known sizes are run alongside the unknown DNA samples so that the sizes of the unknown sample segments can be identified by comparison.

Figure 10.1.
Gel showing two DNA samples plus standards.

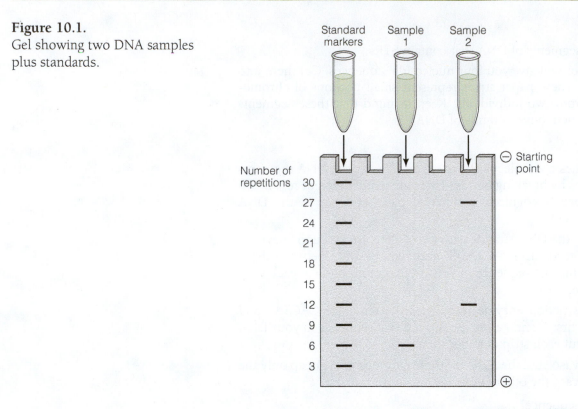

Each band for Sample 1 and Sample 2 in Figure 10.1 represents an allele for this STR locus. How many different alleles does Sample 1 have?

How many repetitions are there in Sample 1's allele(s)?

How many different alleles does Sample 2 have? How many repetitions are there in the allele(s)?

Which sample is homozygous?

5. On a blank sheet of paper, label a lane for the DNA from Person A and another lane for the DNA from Person B. Label a third lane with standard markers ranging up to 25 repetitions. (Remember that the largest standard will be closest to the starting point.)

6. Arrange the STRs you "isolated" in the appropriate places on your paper "gel" and secure them with tape. This is the DNA profile or fingerprint.

What is Person A's genotype at this locus?

What is Person B's genotype at this locus?

Activity B: Application of STR Analysis to Paternity Testing

STR analysis has many different applications, including missing persons investigations, matching a suspect with evidence in criminal cases, wildlife crimes, determining the zygosity of twins, and identifying human remains at disaster sites such as the World Trade Center. In all of these applications, the general objective is to determine the probability that one sample matches another (see Activity C). STR analysis coupled with the laws of Mendelian inheritance is also used in paternity and pedigree testing.

STR alleles are inherited just like the more familiar genes discussed in Lab Topic 9—for example, the genes for tongue-rolling, widow's peak, and astigmatism. Although there are only two known alleles at the "tongue-rolling" locus and there are numerous STR alleles at a given locus, each individual still only has two alleles. One allele was inherited from the mother and one was inherited from the father. As an example, let's use a hypothetical STR locus, STR1. Suppose a man's genotype at this locus is 6,6 and his wife's genotype is 12,27. Use a Punnett square to show the combinations of alleles possible in their children.

What are the possible genotypes of their children, and what is the probability of each genotype?

Use the gel in Figure 10.2 to show the STR analysis of the parents and two children for this locus. Give each child a different genotype.

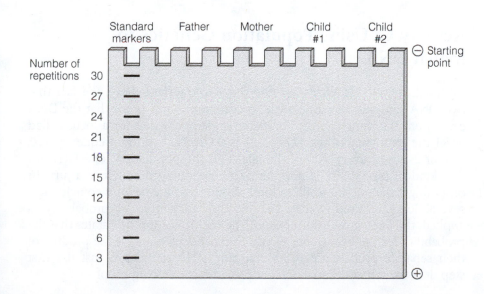

Figure 10.2.
Gel for STR analysis.

Notice that each child gets one allele from each of the parents, but none of the children has a genotype identical to the parents. This type of information is useful in determining paternity and other relationships. Figure 10.3 shows a simple example of how STR analysis can be useful in a paternity case.

Figure 10.3.
STR analysis of paternity case.

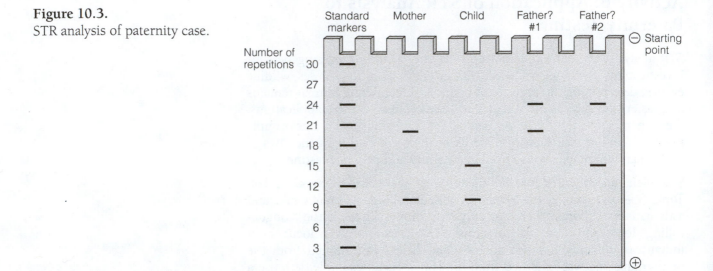

Which man is excluded as a possible father? Explain.

What about the other man? Is he the child's father?

Activity C: Using Population Genetics to Match DNA

Although many STR alleles are rare (have a low frequency), the fact that two DNA samples have one allele in common is not proof that the DNA came from the same source. For example, suppose that for a locus called STR1 the frequency of the "27" allele is 0.02. That's a low frequency, just 2%, or 2 out of every 100 of the alleles in the population for STR1. But that kind of probability is not going to be enough to convince a jury. In order to get the one-in-a-billion kind of odds that are beyond coincidence, several loci are used. When the multiplication rule of probability is applied, the odds grow enormously. (The multiplication rule states that the probability for the occurrence of two independent events is the product of their separate probabilities.) So analyzing DNA samples is just the first step. It is the statistical analysis that provides convincing proof.

The first ingredient needed for the statistical analysis of DNA is the frequencies for the alleles that are to be included in the analysis. Allele frequencies vary by ethnicity and geographic area, so scientists have established databases for a variety of populations. Forensic DNA samples can then be compared to a relatively narrow group of people. By using several different loci, conclusive identifications can be obtained. For example, there are DNA databases that include allele frequencies for U.S. Caucasians, U.S. African-Americans, U.S. Mexican-Americans, U.S. Asians, German Caucasians, U.K. Caucasians, and a few dozen other groups. Some categories are further subdivided by region, such as U.S. Southeast Hispanic and U.S. Southwest Hispanic. Statistical analysis is then based on these known allele frequencies. There are also databases for animals such as cats, dogs, horses, and primates other than humans.

For example, suppose an unidentified body has been found. The general description of the body is similar to those of two missing persons. The first step is to compare the genetic makeup of the missing persons with a DNA sample from the body. DNA for the missing persons may be obtained from the roots of hair left in their hairbrushes or even cells clinging to their toothbrushes. (Lacking a direct sample, DNA from close relatives could be used, but that's more complicated so we'll assume good samples can be taken from the possessions of the missing persons.) The second step is to apply statistics to determine whether the DNA from the body matches the DNA from each of the missing persons. Both missing persons are U.S. Caucasians, so that is the database used for the allele frequencies. Table 10.1 compares the DNA profile of the two missing persons to that of the body that was found.

Table 10.1
Results for 4 STR Loci on DNA Samples from Body and Two
Missing Persons

Locus	Allele	Frequency in U.S. Caucasians	Present in Body	Present in Missing Person 1	Present in Missing Person 2
STR1	9	0.15	X	X	X
STR2	10	0.13	X	X	
STR2	6	0.28			X
STR3	12	0.06	X	X	X
STR3	9	0.02	X	X	
STR3	4	0.13			X
STR4	17	0.14			X
STR4	6	0.04	X	X	

Keep in mind that each individual has two chromosomes containing the same locus. If only one allele is detected, then the individual is homozygous at that locus.

STR locus 1 is represented here by just one allele, "9," and the same allele is present in all three samples. What is the genotype of

The body—

Missing Person 1—

Missing Person 2—

For homozygotes, genotype frequency or probability can be determined by multiplying the allele frequencies. (This follows the rule of multiplication. The probability of receiving a "9" allele from one parent is not affected by the allele contributed by the other parent.)

The frequency for the "9" allele in the U.S. Caucasian population is 0.15. What is the probability that an individual is homozygous for this allele?

Is STR1 helpful in determining the body's identity?

These DNA samples have two alleles at STR locus 2, 10, and 6. What is the genotype of

The body—

Missing Person 1—

Missing Person 2—

Is STR2 helpful in determining the body's identity?

The probability that a U.S. Caucasian has the "10" allele for STR2 is 0.13. What is the frequency of individuals that are homozygous for this allele?

STR1 and STR2 are inherited independently, so the probability of having a combination of genotypes is also determined by the multiplication rule. That is, the probability of a combination of genotypes equals the product of their separate probabilities.

What is the probability that an individual is homozygous both for the "10" allele at STR2 and the "9" allele at STR1?

State the probability in a different way: In a population of 10,000 people, how many would be expected to be homozygous for both of these alleles?

There are 3 alleles at the STR3 locus: 12, 9, and 4. What is the genotype of

The body—

Missing Person 1—

Missing Person 2—

Is STR3 helpful in determining the body's identity?

The probability calculation for heterozygotes is 2 (frequency of allele A × frequency of allele B). Referring to a Punnett square for a heterozygote cross may help you understand why. There are two ways to produce a heterozygous offspring. Let's say the STR3 locus is on chromosome 1. The "12" allele could be on the maternal chromosome 1 and the "9" allele on the paternal chromosome 1, or vice versa. In other words, the genotype could be 12,9 or 9,12. The heterozygote appears twice in the Punnett square.

What is the probability of having the alleles 12 and 9 at STR3? (That is, what is the genotypic frequency?)

These DNA samples have two alleles at STR locus 4, 17, and 6. What is the genotype of

The body—

Missing Person 1—

Missing Person 2—

What is the probability that an individual is homozygous for the "6" allele?

It appears from the comparison of alleles in Table 10.1 that the unidentified body is Missing Person 1. How sure can we be that the body is in fact Missing Person 1 and not some other, unknown, missing person? Calculate the probability of an individual having this genotype (including all 4 loci).

State the probability in a different way: In a population of 1,000,000,000 (one billion) how many people would be expected to match the body's genotype at these four loci? (In other words, what is the probability of getting a match by chance?)

The FBI uses 13 core STR loci, which enables investigators to show that the odds of two DNA samples sharing the same genotype by chance alone are vanishingly small.

E X E R C I S E   1 0 . 2
Application of DNA Profiling

Objective

After completing this exercise, you should be able to

1. Obtain "DNA profiles" and interpret and apply the results.

Now that you have learned the fundamentals of interpreting STRs and calculating probabilities, you will use these techniques to solve a mystery. Your instructor will divide you into investigative teams. Use the background information from the newspaper articles to develop hypotheses about the situation. Your instructor will give you DNA samples to analyze to test your hypotheses. When everyone has finished, each team will present evidence in support of its conclusions.

Procedure

Who fears the revelation of dark secrets from the past? Who sent the anonymous letters? Who is Mr. X and how did he end up in a coma? From the information given in the news articles, formulate hypotheses about the people and events described.

How will DNA profiles help you test these hypotheses?

Your instructor will give you "DNA samples" from a variety of sources. Analyze them and record your results in Tables 10.2 and 10.3. Then answer the questions below, using the allele frequencies given in Tables 10.4 and 10.5 to calculate probabilities to support your conclusions.

On the basis of structural characteristics, detectives decided that the cat hair in the letters and on Mr. X came from one of three breeds: Persian, Turkish Angora, and Birman. To calculate probability, you can assume that police are correct in identifying these three as the only possible breeds.

Grandview Gazette

The Unofficial Word from Grandview Arts and Technical College

Assault Victim Remains in Coma

A man who was found unconscious in the cemetery at the edge of campus remains unidentified and in a coma at this time. Police continue to seek witnesses who may have seen or spoken to this man before the assault. Mr. X, as the hospital staff call him, is a Caucasian possibly in his late 20s or early 30s, 5′10″ tall and weighing about 170 pounds. He has brown hair and eyes and a full beard. Because of facial injuries he apparently suffered in a fight, police don't believe that releasing a current photo would produce any leads. In their investigation to date, the only local connection police have found was a newspaper clipping in the man's pocket describing the drug patent obtained by GATC Professor Jack Watkins a few years ago. Dr. Watkins was unable to identify him. No identification was found in Mr. X's wallet, which contained the stub of a bus ticket from Saskatoon, Canada, a faded photograph of a young woman, a receipt from a private detective agency, and a small amount of money. The Royal Canadian Mounted Police in Saskatchewan have been contacted and inquiries are under way.

The circumstances of Mr. X's injuries are as mysterious as his identity. Since several cigarette ends were found at the scene, police believe that Mr. X and his assailant had a conversation before the fight. They have concluded that a knife found beneath Mr. X belonged to him, since only his fingerprints were on it. There was fresh blood on the knife blade, but no cuts or stab wounds were found on Mr. X's body. Police further theorize that Mr. X's severe head injury, which resulted in the coma, happened when he fell and hit his head on a rock during the scuffle.

One promising lead is the cat hair that was found on his clothing. So far, police have determined that the hair is from a long-haired breed and have identified some possibilities to test. If the hair proves to be from an unusual breed, police may be able to trace the man's identity through the breed registry.

GATC People in the News

More Honors for Professor Watkins

Dr. Jack Watkins, Professor of Biology, has been named a Grandview Distinguished Professor, the highest accolade that Grandview Arts and Technical College offers. The award recognizes faculty for outstanding scholarly contributions to the College.

Dr. Watkins's career at GATC has indeed been illustrious. He has received three teaching awards. In 1999 he earned a patent—and the thanks of a grateful nation!—for a drug that cures the common cold. With characteristic generosity, he has donated some of his patent income to the scholarship fund here at GATC. His alma mater, Great Plains University in Colorado, where he also spent several years as a faculty member, has also benefited from his philanthropy. A lifelong bachelor dedicated to his work, Dr. Watkins carried out his groundbreaking research in a home laboratory (continued on p. 2).

Creek for President?

As GATC searches for a new President, it may find the best candidate in its own backyard. Dr. Diane Creek, who has served GATC as Dean of Students for the last six years, is a finalist for the position. A graduate of Granite College and Midwestern University, Dean Creek held faculty and administrative positions at Great Plains University, Saskatchewan Provincial College, and Southwestern State University before coming to GATC.

To find out more about our Dean of Students and possible President-to-be, the *Gazette* interviewed Dean Creek recently. We met in her campus office, which she has decorated with unique homey touches. Pictures of her husband Jay, who owns a chain of pet stores, son David, a member of the Grandview police department, daughter Mary, a veterinarian, and son Alex, who works with his father, adorn her desk. Given her family's occupations, perhaps it is not surprising that there is also a photo gallery of cats. Dr. Creek explained that she breeds and shows Persian cats as a hobby. She pointed out Cookie, a silky silver cat with startlingly luminous copper eyes whose official name is Best Cook's Star Rainbow Flower (continued on p. 3).

GATC Faculty Targeted by Letters

A STORM OF MYSTERIOUS anonymous letters has blown into the sleepy town of Grandview. For the past several months, rumors have circulated that two well-known campus figures have been the targets of sinister letter-writing campaigns. For the first time, police have publicly acknowledged that the recipients are popular GATC biology professor Jack Watkins and Diane Creek, GATC's Dean of Students.

Dean Creek's administrative assistant alerted police to the first suspicious letter, which arrived while the Dean was out of town. She has received two subsequent letters, one at her campus office and one at her home address. Police have refused to comment on the content of the letters. When the *Gazette* attempted to ask her about them, she would only say "The matter is in police hands. I have no comment." Anonymous sources in the GATC administrative offices, however, have revealed to the *Gazette* that Dean Creek is extremely upset that the police are involved. The administrative assistant who gave the first letter to authorities was fired two days later, allegedly for chronic tardiness. Dean Creek also refused to comment about the firing, citing confidentiality of personnel issues and her former assistant's pending grievance.

Dr. Watkins was willing to talk about the letters he received, but his description of the contents sheds little light on the subject. "Cryptic," he declared, "totally mystifying to me. I wouldn't have bothered the police with them if I hadn't heard rumors about other anonymous letters." However, the police are not yet sure whether the same person is responsible for all of the letters.

The letters themselves have offered few clues to police sources, they say. According to police, they were all printed by computer on plain paper and mailed in ordinary, inexpensive envelopes. They all had different postmarks, leading police to hypothesize that they were sent through a mail-forwarding service. None of the letters bears fingerprints other than those of people who handled the letters locally. However, there is some forensic evidence that will be useful when a suspect is identified. The state crime lab may be able to obtain DNA from saliva on the envelope seals and stamps. In addition, one of the letters to Dr. Creek had a few strands of human hair stuck to the envelope where its glue had gotten damp. Since this letter was received at her home, testing is under way to exclude the recipient and her family as sources of the hair. Some of the letters also contained cat hairs.

Table 10.2

Analysis of Cat DNA. For each STR locus, enter the genotype of each DNA sample.

STR Locus	Cat Hair on Mr. X	Cat Hair in Letter
1		
2		
3		
4		
5		

Table 10.3

Analysis of Human DNA Samples. For each STR locus, enter the genotype of each DNA sample.

STR Locus	Mr. X	Blood on Knife	Blood on Rock	Cig. End 1	Cig. End 2	Saliva on Creek Letter	Saliva on Watkins Letter	Hair on Creek Letter	Dean Creek	Mr. Creek	David Creek	Alex Creek	Mary Creek	Prof. Watkins
1														
2														
3														
4														

Table 10.4
Allele Frequencies for Cat STR Loci

STR Locus	Allele	Persian	Turkish Angora	Birman
1	7	0.02	0.07	0.60
	3	0.38	0.15	0.04
2	10	0.06	0.21	0.22
	5	0.13	0.09	0.00
3	23	0.00	0.11	0.35
	13	0.17	0.13	0.03
	3	0.24	0.01	0.07
4	19	0.29	0.04	0.10
	13	0.11	0.03	0.21
	9	0.03	0.00	0.19
	2	0.13	0.27	0.08
5	15	0.04	0.17	0.16
	11	0.19	0.05	0.09
	8	0.06	0.20	0.16

Table 10.5
Allele Frequencies for Human STR Loci

STR Locus	Allele	Frequency in Caucasians*
1	20	0.24
1	16	0.07
2	15	0.03
2	10	0.17
2	6	0.10
3	12	0.12
3	9	0.02
3	4	0.21
4	24	0.13
4	17	0.11
4	11	0.09
4	7	0.04

* Assume that all characters are Caucasian.

Answer each of the following questions and **explain all of the evidence that supports your conclusion.** Where two DNA samples match, calculate the probability that the match occurred by chance.

What kind of cat did the cat hairs in the two samples come from?

What is the probability of the combination of genotypes in the two cat hair samples occurring? That is, out of one billion cats of this breed, how many would you expect to share this genotype?

Whose (human) hair was on the letter sent to the Creek house?

Who wrote the anonymous letters?

Who assaulted Mr. X?

Who is Mr. X?

What does Dean Creek have to hide?

What does Professor Watkins have to do with the case?

Questions for Review

1. The frequencies of many known human alleles are very high compared to those of the STR alleles. For example, the frequency of the i allele in the ABO blood-typing system is 0.67 and the frequency of the A allele is 0.26. Why are the allele frequencies so low for the STR alleles?

2. Explain how the following are used in STR analysis:

    Restriction enzymes—

    Electrophoresis—

3. In matching DNA found at a crime scene to a suspect's DNA, why isn't it enough to show that the DNA samples have the same alleles? Why must statistical analysis be used?

Bacteria and Disease

Introduction

Members of the two prokaryotic domains, Bacteria and Archea, are not visible to the unaided eye. The organization of their cells is simpler than that of the much larger eukaryotic cells. However, prokaryotes display a wide variety of metabolic activities. Some are autotrophic. That is, they can synthesize their own organic compounds using energy from light (photosynthetic) or inorganic chemicals (chemosynthetic). Heterotrophic prokaryotes acquire their energy by feeding on organic molecules. Cellular respiration is also diverse—there are at least a dozen types of fermentation pathways. Because of this diversity, prokaryotes are found in many different habitats. For example, some are important members of soil and aquatic communities and some inhabit the skin and bodies of animals. Despite the fact that the vast majority of prokaryotes cause no harm to humans and some even perform essential functions, most of our experience is with those that cause disease. About half of all human diseases are caused by members of the domain Bacteria, including strep throat, food poisoning, leprosy, tuberculosis, gonorrhea, and even ulcers, to name just a few. In this lab topic you will learn about some aspects of bacteria, especially pathogens (agents of disease).

Outline

Exercise 11.1: General Features of Bacteria

    Activity A: Bacterial Shapes

    Activity B: Colonies

Exercise 11.2: Investigating Microorganisms and Disease

    Activity A: Gram Stain

    Activity B: Antibiotic Sensitivity Test

    Activity C: Antibiotic Resistance

    Activity D: Transmission

E X E R C I S E   1 1 . 1
General Features of Bacteria

Objectives

After completing this exercise, you should be able to

1. Describe how prokaryotic cells differ from eukaryotic cells.
2. Draw and name the three major morphological types of bacteria.
3. Explain how bacterial colonies arise and how they may be useful in classification.
4. Explain what the exposure test tells us about the presence of bacteria in the environment.

Members of the Domain Bacteria are prokaryotic. All other organisms are eukaryotic, including the protistans and fungi that are also considered microorganisms. **Prokaryotes**, as you may recall from studying cells, are much smaller than eukaryotes and lack their complex organization. Chiefly, they do not have the specialized membrane-bound organelles such as mitochondria or chloroplasts that are found in eukaryotes. Prokaryotes do have ribosomes, which function in protein synthesis. Although they have the same function as eukaryotic ribosomes, the ribosomes of prokaryotes are somewhat different in composition. These differences are useful for developing antibacterial drugs: Prokaryotic ribosomes can be targeted while eukaryotic ribosomes are unharmed.

Most bacteria have a cell wall. Many are also surrounded by some sort of capsule of gelatinous material, which may play a role in pathogenicity. For example, in the pneumonia-causing bacterium, forms that have a capsule have the ability to cause the disease, while those without a capsule do not.

In using drugs to combat pathogens, scientists want to select chemicals that interfere with some process or part of the bacterial cell without harming the host (human) cells. For example, if a drug that inhibits glycolysis were administered to a patient, it would damage his or her own cell function as well as killing the bacteria. Considering the differences between bacteria and animal cells discussed above, what aspects of bacterial structure might be targeted?

Activity A: Bacterial Shapes

Most bacteria have one of the three shapes illustrated in Figure 11.1: coccus (spherical), bacillus (rod), or spirillum (spiral). The characteristic shapes help microbiologists classify and identify bacteria.

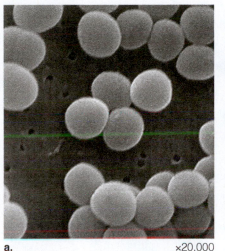
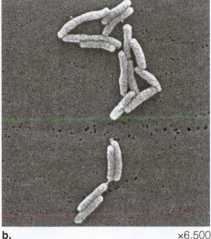
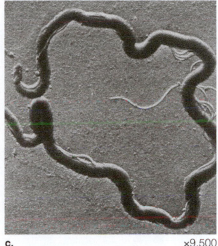

a. ×20,000   b. ×6,500   c. ×9,500

Figure 11.1.
Bacterial shapes. In addition to the coccus (a), bacillus (b), or spirilla (c) shape of an individual bacterium, aggregations of bacteria in clusters or strings may provide an additional characteristic for classification.

In the following procedure you will look at prepared slides in order to distinguish the three bacterial shapes. You will also view a slide that is stained to show the capsule of a bacterium.

Procedure

1. Locate the bacteria first at low power and focus.
2. Change the objective to high power and center several bacteria in the field of view. You may need to adjust the fine focus slightly, but don't adjust the coarse-focus knob.
3. If your microscope is equipped with an oil immersion lens, turn the nosepiece so that no objective is directly over the slide. Place a small drop of immersion oil on the area of the slide that you were viewing. Then turn the oil objective into place and focus. Again, you may need to adjust the fine focus slightly.

4. Check with your instructor to make sure you are looking at the right thing, and then draw the bacteria in the space below.

Coccus   **Spirillum**

Bacillus   **Bacterium**
with capsule

Can you see any internal structures in the bacteria? Explain.

If you were looking at a eukaryotic cell at high power on your microscope, what structures might you see?

 If you used an oil immersion lens, be sure to clean the oil off the lens and the slide with a clean piece of lens paper.

Activity B: Colonies

Bacterial form clearly does not provide much information for classification. Much higher magnification would have to be used to see any characteristic other than shape, though sometimes aggregations (the ways that cocci cluster together) can be informative. One way to "magnify" bacteria is to look at many millions of bacteria together rather than looking at a single cell.

Bacteria may divide as often as once every 10 minutes, so a single bacterium that finds itself in a hospitable environment can give rise to a population of millions in 24 hours. While the individual bacterium is not easily seen, the **colony,** or the population of bacteria derived from one or a few cells, is visible to the unaided eye. The appearance of the colony, such as its color and shape, may be useful in describing the species of bacteria.

Bacteria are found everywhere. Last week, petri dishes of agar, a gel nutrient medium, were exposed to various sources of microorganisms such as the sole of a shoe and a dog's paw. After exposure, the dishes were kept in a favorable environment, with the result that each dish now contains an array of colonies that arose from the bacteria present in the various sources. Fungal spores, single cells that are capable of germinating and growing into colonies of fungus, are also found everywhere, so the petri dishes also contain fungal colonies. In general, the colonies that have a fuzzy appearance are fungi, while the rest are bacteria.

In this exercise, you will compare the diversity of microorganisms from various sources.

If bacteria and fungal spores are found everywhere, why don't we see colonies of bacteria and fungi everywhere?

Procedure

1. Get a petri dish from your instructor and count the number of different types of colonies present. Use a dissecting microscope to examine the colonies more closely. Figure 11.2 should help you distinguish how many different types there are.

 Do not open the petri dishes. They could be harboring allergenic or pathogenic microorganisms.

Figure 11.2.
Characteristics used to differentiate bacterial colonies.

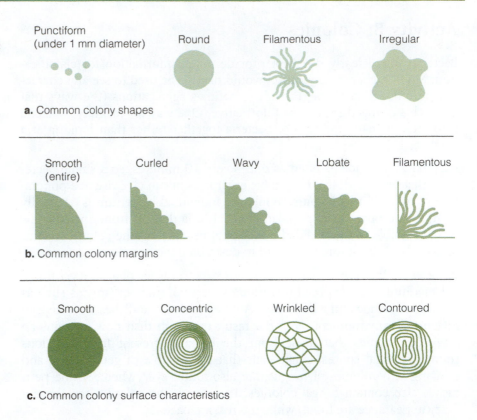

a. Common colony shapes

b. Common colony margins

c. Common colony surface characteristics

2. Record your data in Table 11.1.

3. When you have finished with one dish, trade with another lab team so you can examine dishes from a variety of sources. Record your data in Table 11.1.

4. Compare the diversity of microorganisms found in different sources.

Which source shows the greatest diversity? Which has the least diversity?

Table 11.1
Diversity of Bacterial Colonies from Different Sources

Source	Number of Different Types of Colonies

Which sources have the most similar types of bacteria? Can you explain why?

Can you identify a type that appears in all the petri dishes?

E X E R C I S E 1 1 . 2
Investigating Microorganisms and Disease

Objectives

After completing this exercise, you should be able to

1. Explain what the Gram stain is and why it is useful.
2. Define antibiotic and give examples of how antibiotics work.
3. Explain how the antibiotic sensitivity test works.
4. Explain how the antibiotic sensitivity test could be used to determine the effectiveness of a drug.
5. Explain what antibiotic resistance is, how it arises and spreads, and why it causes serious concern.
6. Explain how diseases can be transmitted, and how to trace an epidemic to its source.

Activity A: Gram Stain

There are two major types of bacterial cell walls that can be differentiated by a procedure known as the Gram stain (named for the person who invented it). This is one of the most useful procedures in bacterial identification. Gram-positive bacteria tend to be more sensitive to most antibiotics than gram-negative bacteria, though there are major exceptions. Some antibiotics, however, work better on gram-negative bacteria. Broad-spectrum antibiotics work on both gram-positive and gram-negative bacteria.

In Activity A, you will perform a Gram stain. In Activity B, you will look at the effect of various antibiotics on cultures of gram-positive and gram-negative bacteria.

 Wear safety glasses while you are performing the following procedure.

Procedure

1. Take a clean slide and make a small circle in the center of it with a grease pencil. Then turn the slide over so the unmarked side is up.

2. Place a very small drop of water in the circle. With the flat end of the toothpick, scrape some tartar from your teeth (your back teeth will probably have the most bacteria). Stir it into the drop of water and spread it over the circled area.

 Dispose of the toothpick in the container provided.

3. Let the slide air dry.

4. Pick up the slide with a clothespin and heat fix it by passing it rapidly through the flame of an alcohol lamp several times. If you don't do this, your bacteria will be washed away at the next step, but you don't want to fry them, either!

⚠️ **Be careful to keep everything but your slide away from the flame.**

5. Place a small petri dish upside down in a large petri dish, which will hold the excess stain. Set the slide across the small petri dish and flood it with crystal violet stain. (See Figure 11.3a). Leave the stain on the slide for one minute.

6. Drain off the crystal violet stain and rinse the slide gently with water from the wash bottle (see Figure 11.3b).

Figure 11.3.
(a) Staining dish. The small petri dish is a platform for the slide; the large dish catches excess stain.
(b) Rinsing the slide. Use a gentle stream of water to rinse stain from the slide.

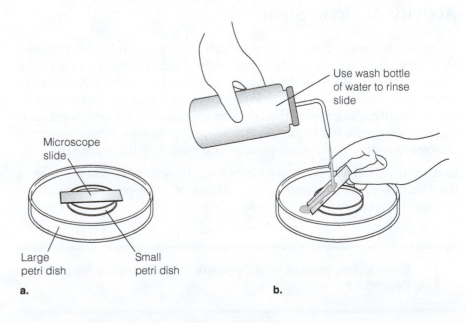

Microscope slide

Use wash bottle of water to rinse slide

Large petri dish

Small petri dish

a.

b.

7. Place several drops of Gram's iodine on the slide and let it sit for 1 minute.

8. Wash the iodine off the slide as in step 6.

9. Hold the slide at an angle and apply 95% alcohol drop by drop until the alcohol running off is clear rather than blue. This should take 10–30 seconds—don't overdo it!

10. Rinse the slide briefly with water. At this point the gram-positive bacteria are stained violet or blue and the gram-negative bacteria are colorless. The next step, called a counterstain, allows you to see the gram-negative bacteria.

11. Flood the slide with safranin, the counterstain, for 30 seconds.

12. Rinse the slide gently with water. Hold the slide upright to let as much water as possible drain off; then let it air dry before you look at it. No coverslip is used on these slides. Gram-positive bacteria will be violet or blue; gram-negative will be red.

13. Draw the bacteria you see in Table 11.2. Record the Gram reaction and structural type for each. A species of *Streptococcus* (gram-positive cocci in chains) is the cause of tooth decay. Did you find any in your plaque?

Table 11.2
Bacteria in Plaque

Bacteria (sketch)	Gram Reaction (positive or negative)	Type

 When you have completed the exercise, dispose of the excess stain and your slide as directed by your instructor.

Activity B: Antibiotic Sensitivity Test

The human immune system is very effective at checking the growth of pathogens that invade the body. However, it has limitations. A pathogen that does manage to reproduce rapidly in the body may overwhelm the immune system's ability to respond, resulting in serious illness or death. Antibiotics, drugs that either kill or inhibit the growth of bacteria, assist the immune system. Once antibiotics bring the population of bacteria under control, the immune system can regain the upper hand and finish off the rest.

When you think about how many times you and your family have taken antibiotics, it is hard to conceive of a world without them. But before the development of penicillin in the mid-1940s, deadly diseases such as bubonic plague, cholera, and syphilis had no cure. Even strep throat was potentially lethal. For example, consider the bacterium that causes tuberculosis (TB). A disease known at least since 1000 BC, TB infects the lungs. Symptoms include a bloody cough, fever, and a gradual wasting away of the body that ends in death. Before antibiotics, TB claimed countless victims, including many famous ones such as the poet John Keats, authors Emily Brontë, Franz Kafka, and George Orwell, and Doc Holliday, legendary figure of America's Old West. The antibiotic streptomycin, introduced in the 1940s, proved to be the miracle cure for tuberculosis. Other effective antibiotics were introduced in the next decade.

As discussed in Exercise 11.1, antibiotics should interfere with pathogens without affecting the cells of the body. Thus, their action usually relies on targeting some aspect of cell structure or function that differs between prokaryote and eukaryotes.

Why would antibacterial agents usually not be effective against fungi?

The antibiotics most familiar to us are antibacterial agents, probably because we are more likely to have had bacterial than fungal diseases. Some of these antibiotics, including erythromycin, tetracyclines, and compounds related to streptomycin (gentamycin, kanamycin, and neomycin) work by binding to bacterial ribosomes. How does this affect bacteria?

Why would eukaryotic organisms not be affected by antibiotics?

Why would antibiotics not be effective against viruses?

Many antibiotics in use today were originally derived from certain types of fungi and bacteria that live in the soil, such as *Penicillium* and *Streptomyces*.

Suggest a reason why *Streptomyces* and *Penicillium* might have evolved the ability to produce antibacterial chemicals.

In this exercise you will determine the effectiveness of different antibiotics in inhibiting the growth of bacteria. Two petri dishes have been prepared with cultures of bacteria, one with a gram-positive species (*Streptococcus aureus*) and one with a gram-negative species (*E. coli*). Small disks soaked in various antibiotics were placed on the dishes. Each disk is coded with the name of the antibiotic; note the key to the antibiotics next to the dishes.

If the bacteria growing on the plate are sensitive to a given antibiotic, their growth will be inhibited and no bacteria will be observed growing near that disk. That is, the antibiotic will cause a **zone of inhibition** as shown in Figure 11.4.

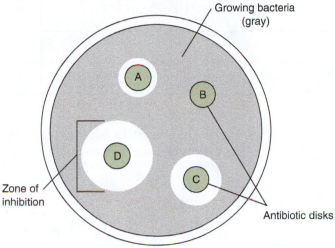

Figure 11.4.
Antibiotic sensitivity test. In this example, antibiotic D is most effective and antibiotic B is least effective against these bacteria.

Procedure

1. Measure in millimeters the diameter of the zone of inhibition for each antibiotic.

 Do not open the petri dishes because they may have been contaminated by pathogenic or allergenic microorganisms.

---

2. Record the results in Table 11.3.

Table 11.3
Results of Antibiotic Sensitivity Test

Antibiotic	Zone of Inhibition (mm)	
	Gram-Negative Bacteria	**Gram-Positive Bacteria**

Do gram-positive and gram-negative bacteria react similarly to the same antibiotic? Explain the results you have recorded above.

How might a medical lab use this test?

Activity C: Antibiotic Resistance

As treatments became available for formerly incurable diseases, the development of antibiotics seemed almost miraculous. It even seemed possible that some pathogens could be wiped out entirely. For example, experts confidently predicted the total eradication of tuberculosis by 2025. Indeed, in industrialized countries where treatment was available, cases of TB became very rare. But in the 1980s, TB cases in the United States as well as around the world showed a marked increase. In 1993, the World Health Organization declared the disease, which kills between two and three million people each year, a global emergency. Why did the health care community swing so rapidly from optimism to alarm over antibiotic use?

Bacteria are organisms that are subject to natural selection by their environment—in this case, the human body. They can evolve. Changing their environment by introducing antibiotics into it creates an opportunity for evolutionary change. Even amid the euphoria of the new era of antibiotics, some microbiologists noticed warning signs of trouble ahead. They discovered that some bacteria harbored a natural resistance to penicillin. That is, the bacteria had some genetically determined mechanism that enabled them to escape the drug's effect.

Suggest a reason why some bacteria might naturally carry a gene that defeats an antibiotic. (Hint: Refer to Activity B. Why does the *Penicillium* fungus produce the antibiotic?)

It takes a large population of bacteria to produce disease. For example, the lungs of a person who has an active case of tuberculosis contain hundreds of millions of bacteria. Those bacteria are not genetically identical to each other, and as a result they also vary phenotypically. The phenotypic differences may include an ability to circumvent the effects of antibiotics. Some bacteria may produce an enzyme that degrades or deactivates an antibiotic. Other bacteria may have the ability to prevent a drug from entering their cells, or they may be able to disguise the drug's molecular target, or even to expel the antibiotic before it has a chance to work.

The different writing implements provided for this activity simulate the genetic and phenotypic differences in the bacterial population.

Procedure

1. Use the pens and pencils provided to fill the diagram of lungs below with dots or circles representing bacteria.

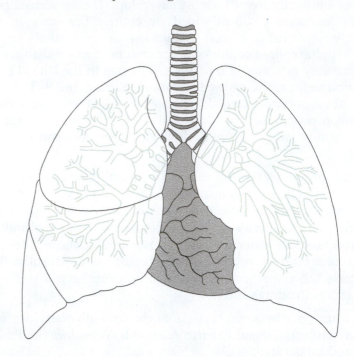

2. The erasers represent antibiotics. Your instructor will tell you which one to "administer" first. Rub very lightly at first over the entire area, then use increasing pressure. But don't bear down so hard that you tear the paper!

Describe the results, including which bacterial types have been completely erased, which are fainter but still present, and which are as bold as ever.

When you take an antibiotic, it begins killing some bacteria immediately. But some are harder to kill than others—it takes a higher concentration of the drug to do the job. When you don't finish your antibiotic prescription, you leave a population of the tougher bacteria unscathed. This gives the bacterial population a chance to resume rapid reproduction. As a result, your immune system may be overwhelmed again, causing a recurrence of your illness.

3. Repeat step 2 using the second "antibiotic" designated by your instructor.

Describe the results, including which bacterial types have been completely erased, which are fainter but still present, and which are as bold as ever.

4. Repeat step 2 using the third "antibiotic" designated by your instructor.

Describe the results, including which bacterial types have been completely erased, which are fainter but still present, and which are as bold as ever.

5. Fill the lung diagram with "bacteria" that have survived all of the antibiotic treatments.

When this individual passes the disease on to someone else, which bacteria will the newly infected person acquire?

In the process of natural selection, members of the population that are better suited to their environment are more likely to survive and reproduce. When the environment—in this case, the human body—changes, the phenotype that is best adapted to the environment changes. The resistant bacteria reproduce rapidly, so within a short period of time, the person is once again suffering from a full-blown infection.

Strains of the tuberculosis bacterium that are resistant to multiple antibiotics have emerged since the 1990s. These strains are more difficult and costly to combat (almost a billion dollars were spent controlling an outbreak in New York City in the early 1990s). But that is just one example of a pathogen that has developed antibiotic resistance. Others include bacteria that can cause pneumonia, toxic shock, gonorrhea, food poisoning, internal infections, and (parents, take note!) childhood ear infections. To make the situation even more alarming, the resistance genes are carried on plasmids, small circles of DNA that can be readily passed from one *species* of bacteria to another.

Activity D: Transmission

Exercise 11.1 showed that microorganisms are found in many different places, but that some are more likely to be found in one place than another. Pathogens are most likely to be found on or in the vicinity of their human hosts, where they can easily be transmitted to new hosts. For example, tuberculosis is highly contagious. From its breeding grounds in the lungs, it spreads by coughing, sneezing, spitting, or even talking. It only takes a small number of inhaled bacteria to start an infection. Other pathogens may be transmitted by direct contact (including sexual contact), by contaminated objects, or by excretory products and body fluids. Animal carriers are responsible for some disease transmission. In addition, certain diseases can be transmitted from mother to fetus in the uterus or during birth.

Many times when we are ill we have no idea how we contracted the infection. If there is a local outbreak of a disease, though, we are aware that something is "going around" and can often identify "whose" cold we got. For example, small epidemics on college campuses frequently get started at the beginning of the semester or after a vacation. A student returns to school with a cold he got from his little sister and passes it on to his roommate, who transmits it to six people in his biology lab, who take it back to their roommates, and so on.

Epidemics of some diseases are cause for serious concern, and health departments work to identify the transmission routes. This may involve determining common factors among the victims (for example, in an outbreak of food poisoning) or tracing all contacts of the victims (for example, in sexually transmitted diseases). In this exercise, you will simulate transmission of an infection and trace its course through the class.

Procedure

1. Each student has a stock bottle of solution. One student's solution is different from all the rest, but the difference is not visible.

2. Transfer 3 pipetsful of solution from your stock bottle to a clean test tube, using a pasteur pipet. You can think of this as the aerosol droplets from a sneeze or as the exchange of bodily fluids in a sexual contact.

3. Find a person at random in the class and exchange a dropperful of solution with him or her. That is, you put a dropperful of your solution into that person's test tube, who then puts a dropperful of his/her solution into your test tube.

4. Record the name of your contact below.

5. When your instructor says to, repeat steps 3 and 4 twice more, so you have contacted three *different* people.

   List your contacts.

   First contact: _____

   Second contact: _____

   Third contact: _____

6. Put a dropperful of phenol red into your test tube and record the color: _____

   Phenol red is an acid/base indicator: Acidic solutions turn yellow, and basic solutions turn red when phenol red is added. The "infected" solution was basic, and all of the others started out acidic. The solutions in the test tubes of "infected" individuals will now be basic, or red.

7. Record in Table 11.4 the names and contacts of infected individuals in the class.

8. Determine the transmission routes and answer the following questions.

What was the *maximum* number of people in your lab class who could be infected after each round of contacts?

First round: _____

Second round: _____

Third round: _____

Table 11.4
Record of Contacts of "Infected" Persons

Infected Person	Contacts		
	First	Second	Third

Explain your answers.

After the three rounds of contacts, what would be your chances of finding an uninfected partner? How would your chances change after a fourth round?

Why might you find that fewer than the maximum number were infected?

How might the results have been different if you had been given 5 minutes and were told to make as many contacts as you wanted?

In this lab exercise you could not avoid contacting an infected individual because you couldn't tell who was infected until it was too late. What are some actual diseases where this might be the case?

Questions for Review

1. Name several different structures that eukaryotes (for example, animal cells) have that prokaryotes (for example, bacteria) do not have.

2. How and under what conditions do bacterial colonies arise?

3. Smoking irritates the cells lining the respiratory tract, causing them to produce more mucus than usual. How is this related to the fact that smokers have more respiratory infections than do nonsmokers?

4. What is an antibiotic?

5. Give the most common examples of how antibiotics work.

6. What part of the bacterial cell is stained by the Gram stain? Why is this procedure useful?

7. When an antibiotic binds to bacterial ribosomes, how (specifically) does that affect the bacterium?

8. Explain how the antibiotic sensitivity test works.

9. How could the antibiotic sensitivity test be used to determine whether a particular drug is effective in treating a particular bacterial disease?

10. Explain how the process of natural selection results in populations of bacteria that are not harmed by antibiotics.

11. List several ways that disease can be transmitted.

12. Using the exercise you did in lab as an example, explain how an epidemic can be traced to its source.

13. What is another strategy epidemiologists might use to determine the source of a disease?

Plant Diversity

Introduction

The plant kingdom arose from aquatic algal ancestors more than 400 million years ago. A survey of the kingdom is a study in the evolutionary history of plants as they successfully made the transition from water to land. Figure 12.1, on the next page, shows the phylogenetic relationships among the four major plant groups you will study in this lab topic.

As we describe the divisions we will note a series of anatomical adaptations for living on land, including:

- Vascular tissue, which is specialized for transporting water and nutrients inside the plant.

- A root system, which anchors the plant and absorbs minerals and water from the soil.

- A cuticle and/or cork layer on above-ground parts, which retards water loss.

- The presence of pores in the above-ground parts that allow for gas exchange.

Most of this lab topic, however, will focus on the other major evolutionary trend: the ability of plants to reproduce on land. In the earliest plants, sperm were required to swim to the egg in order to achieve fertilization and thus needed to remain in or near water. For plant reproduction to be completely successful on land, some other means of bringing the gametes together was necessary. Pollen, which brings the sperm-producing cell to the egg-bearing female structure via air, solved this problem.

The evolution of the seed, which allowed plants to provide their offspring with a good start in the hostile terrestrial environment, was another critical development.

In this lab topic you will study four representative life cycles—moss, fern, pine, and flowering plant—and see how each group's reproductive adaptations contributed to its success. You will also have an opportunity to see living representatives of several other groups of plants.

Outline

Exercise 12.1: The Alternation of Generations Life Cycle
Exercise 12.2: Seedless Nonvascular Plants
Exercise 12.3: Seedless Vascular Plants
    Activity A: Club Mosses
    Activity B: Horsetails
    Activity C: Ferns

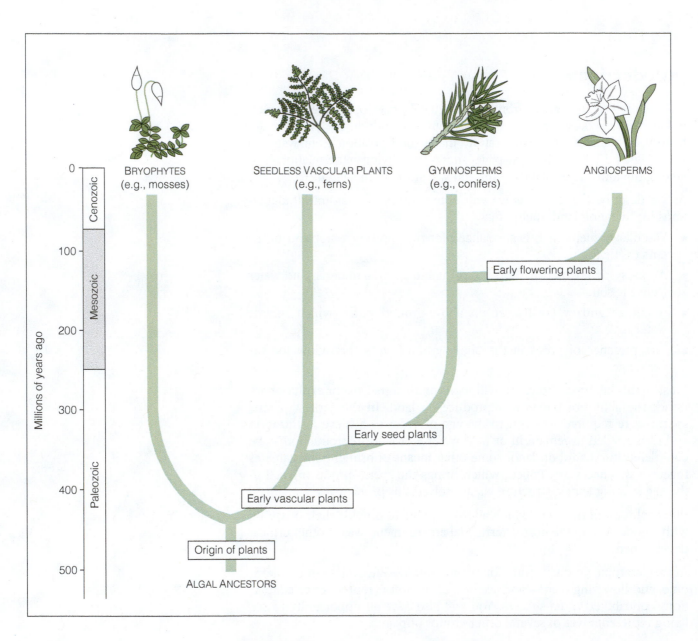

Figure 12.1.
Phylogenetic tree of major plant groups. The bryophytes were the earliest land plants and are represented in modern times by the mosses. The first vascular plants were seedless and gave rise to seed plants during the Paleozoic era, over 300 million years ago. The most recent addition to the plant kingdom is the angiosperms, or flowering plants.

E X E R C I S E   1 2 . 1
The Alternation of Generations Life Cycle

Objective

After completing this exercise, you should be able to

1. Explain the general pattern for the alternation of generations life cycle.

The process of sexual reproduction involves **meiosis**, production of haploid cells that contain only one set of chromosomes (**1n**). Meiosis is followed by **fertilization**, restoration of diploid cells, which have two sets of chromosomes (**2n**), by fusion of haploid cells. In animals such as ourselves the process is familiar and straightforward. Our diploid bodies have specialized cells that undergo meiosis to produce eggs or sperm, which are called gametes. The gametes join during fertilization to form the diploid zygote from which a new individual develops.

Plants, however, go through an **alternation of generations**, as shown in Figure 12.2. When a diploid plant undergoes meiosis, it produces haploid **spores**, not gametes. Spores are made in a specialized structure called a **sporangium**. (The suffix -angium is from a Greek word that means "container.") The spores undergo mitotic divisions to produce a plant that is entirely composed of haploid cells.

Cells in the **gametangium**, a specialized structure in the haploid plant, undergo mitosis to produce gametes. Male and female gametes (sperm and egg) must then be brought together during fertilization. Animal sperm, as you probably know, are motile: They swim through fluid to reach the egg. Some plant sperm are also motile, but if this is the case then the plant requires a film of water for reproduction. More advanced plants have evolved sperm that do not require water for transport, as you will see in this lab. In any case, when the sperm fertilizes the egg, a zygote is formed. The zygote grows by mitotic cell divisions into a diploid plant.

Because the diploid plant produces spores, it is called a **sporophyte.** The haploid plant, which produces gametes, is known as a **gametophyte.** In some groups of plants the sporophyte and gametophyte are both autotrophic, or capable of photosynthesis, and thus are able to live independently of each other. In other plants, one generation is dependent on the other for nutrition—that is, one generation is heterotrophic.

What characteristic of a plant indicates that it is autotrophic? (Hint: What is necessary for photosynthesis?)

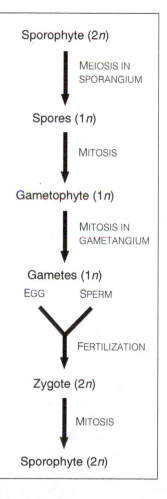

Figure 12.2.
Alternation of generations life cycle.

The alternation of generations is shown in Figure 12.2 as a linear progression through time because each sporophyte gives rise to a gametophyte, which then gives rise to a *new* sporophyte. However, we could just as easily trace a life cycle from gametophyte to gametophyte, and in fact this is done when the gametophyte phase of the life cycle is dominant.

Diagrams of life cycles are often drawn as circles to emphasize that you can start at any point in the life cycle and follow it through. You should learn the alternation of generations pattern thoroughly so that you can understand a life cycle regardless of the format used to present it.

On Figure 12.3, fill in the blanks to indicate where meiosis, mitosis, and fertilization (events) occur and where the spores and zygote (structures) are found in the life cycle.

Figure 12.3.
Review of the structures and events in the alternation of generations life cycle.

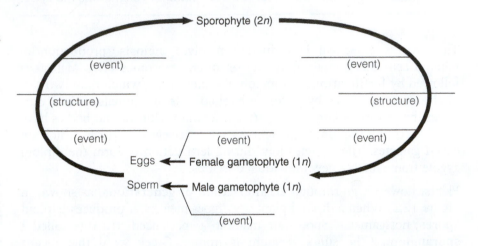

In this laboratory you will view specimens representing the different stages of each life cycle. A diagram of these structures in the lab manual will help you fit the pieces together and see the entire life cycle. For each life cycle, you will be asked to identify events (meiosis, mitosis, and fertilization), whether the tissue is haploid or diploid, and certain structures, including the gametophyte, gametes, zygote, sporophyte, and spores.

You should approach each life cycle like a puzzle. First, locate something you recognize—for example, fertilization. Once you do that, you know what happens next in the general life cycle, and you can deduce the rest from that.

Keep in mind that:

- Only diploid cells can undergo meiosis.
- Meiosis occurs only once in the life cycle.
- Gametes are produced by mitosis.
- Fertilization is necessary to restore diploidy.

The diagrams in the lab manual show the life cycles as circular, indicating that the same events are repeated in each generation. It is more convenient in the laboratory, however, to map out each life cycle display in a linear manner. The beginning point is either a gametophyte or a sporophyte (whichever is dominant) and the end point is a *new* gametophyte or sporophyte.

EXERCISE 12.2
Seedless Nonvascular Plants

Objective

After completing this exercise, you should be able to

1. Identify in living materials or on a diagram all of the structures and events in the moss life cycle.

The **bryophytes** (Greek *bryon*, moss) are relatively small plants that live in habitats where water is available at least during part of the year, since water is necessary for reproduction. They have been called the amphibians of the plant kingdom because amphibians (frogs, toads, and salamanders) have a similar requirement for water. There are 24,000 species of bryophytes. They are most common in tropical and subtropical regions but are found in a wide variety of habitats, including deserts. Since they are extremely sensitive to air pollution, their absence from otherwise suitable habitats is an indicator of pollution.

The descriptions of plants given here refer to the gametophyte phase of the life cycle, since that is predominant. Most bryophytes lack **vascular tissue,** the specialized food- and water-conducting tissue found in the higher plants. They do not have true stems, leaves, or roots, which places severe limitations on their size. Nevertheless, botanists use the terms leaf and stem to refer to their leaflike and stemlike structures. The rootlike organs are called rhizoids (Greek *rhiza*, root; *oid,* like).

You are probably already familiar with mosses, which are leafy and upright. The bryophytes also include a group of plants called liverworts. (*Wort* is part of numerous plant names; it is derived from *wyrt,* an Old English word meaning "plant.") Some liverworts are leafy, but many are flat (sort of liver-shaped) and undifferentiated into leaves and stems, like the specimen in the lab. Another liverwort is shown on Color Plate 2.

Procedure

1. Examine the liverwort under the dissecting microscope and sketch it in the margin of your lab manual.

   Notice the pores in the liverwort's surface. What do you suppose is the function of the pores?

   Having open pores is a problem for land plants, since water the plant needs is readily lost through the pores. Most plants have **stomata** instead. Stomata are special pores that can be closed off by the surrounding cells. Some bryophytes do have stomata, but they generally remain open all day so they are not really effective at preventing water loss. Higher plants open and close their stomata to balance carbon dioxide intake with water loss.

2. Look at the moss life cycle in Figure 12.4, which represents the bryophytes. Label the arrows that indicate where mitosis, meiosis, and fertilization occur (mitosis occurs more than once).

3. Observe the following structures in the living material and label them on Figure 12.4:

Female gametophyte	Zygote
Male gametophyte	Embryo
Egg (female gamete)	Sporophyte
Sperm (male gamete)	Spores

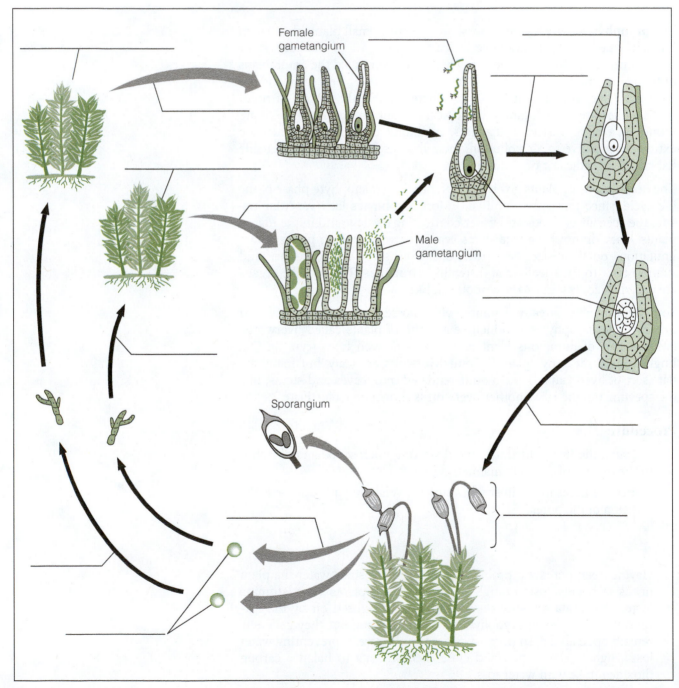

Figure 12.4.
Moss life cycle.

4. With a colored pencil, color all of the **haploid** structures in Figure 12.4.

   How does the sperm get to the egg?

   How does the gametophyte obtain its food (is it autotrophic or heterotrophic)?

   How does the sporophyte obtain its food (is it autotrophic or heterotrophic)?

   Which phase of the life cycle is larger and lives longer?

E X E R C I S E   1 2 . 3
Seedless Vascular Plants

Objectives

After completing this exercise, you should be able to

1. Identify representative members of the three divisions of seedless vascular plants.
2. Identify in living material or on a diagram all of the structures and events in the fern life cycle.

In vascular plants we see increasing refinements of the adaptations for life on land. As the name implies, members of this group have specialized vascular tissue consisting of xylem and phloem for the transport of water and food. Cuticles, which are fatty or waxy layers, are a more sophisticated means of protecting leaves and stems from disease-causing organisms and desiccation. The spores of these plants are also more protected from drying out than are the spores of bryophytes.

Seedless vascular plants make up only a small part of today's plant kingdom, but during the Carboniferous period, 360–290 million years ago, seedless forests dominated the landscape. Some of the groups that have living representatives are listed below, but most members of the seedless vascular plants are extinct.

Use the margin of your lab manual to sketch the specimens that are on display in the lab.

Activity A: Club Mosses

The five living groups of club mosses, which include 1,000 species, have very small leaves and roots arising from an underground stem. The largest genus is *Selaginella;* one of the 700 species of *Selaginella* is on display in the lab. The living club mosses are all herbaceous; that is, they never develop wood. Of the extinct forms, some were woody and treelike and were dominant plants of the great Carboniferous forests, which formed the extensive coal deposits that are mined today.

Activity B: Horsetails

Horsetails were also dominant plants of the coal-forming age. Today, however, only one genus, *Equisetum,* survives. One of the 15 species is illustrated on Color Plate 2. Because of siliceous deposits in the epidermal cells, horsetails have a sandy, abrasive texture and were used in colonial and frontier times to scour pots and pans. Feel the roughness of the stem of the specimen in the lab. Horsetails are quite common in moist areas. Look for them near streams and ditches.

Activity C: Ferns

Although we are familiar with most of the 12,000 species of ferns as woodland plants, there are also aquatic and even desert forms. In size they range from a water fern with leaves 2 cm long to the giant tree fern, which may have leaves up to 5 m in length. The roots of ferns originate from underground stems. The leaves, called fronds, also arise from the underground stems, and they have more vascularization than the other seedless vascular plants. In most species, spores are produced in rusty patches called sori on the underside of a leaf or in large clusters on a leaf that is specially modified for reproduction. Because sori give the frond a spotted appearance, many people mistake these reproductive structures for some type of disease.

Procedure

1. On Figure 12.5, label the arrows that indicate where mitosis, meiosis, and fertilization occur (mitosis occurs more than once).
2. Observe the following structures in the living material on display and label them on Figure 12.5:

Sporophyte	Egg (female gamete)
Spores	Sperm (male gamete)
Gametophyte	Zygote

3. With a colored pencil, color all of the haploid structures in Figure 12.5.

How does the sperm get to the egg?

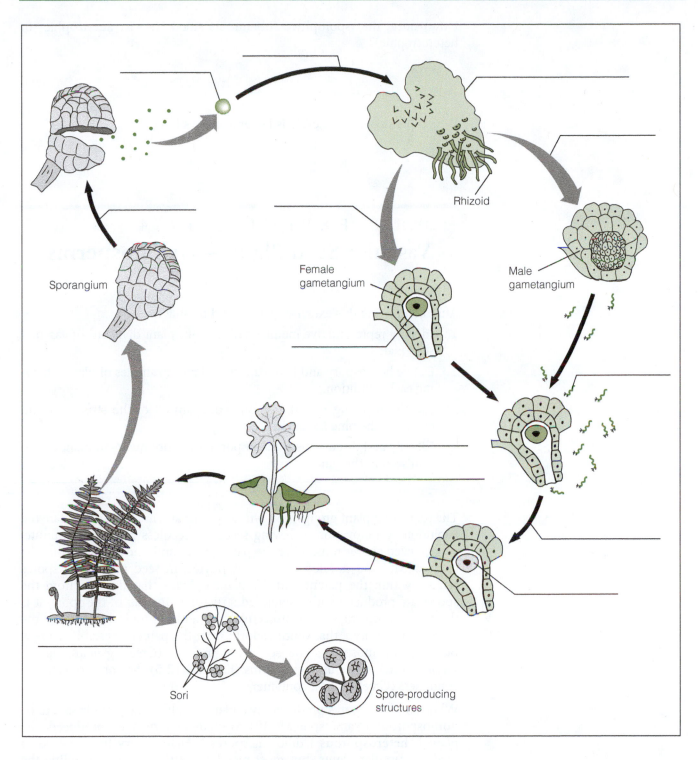

Figure 12.5.
Fern life cycle.

How does the gametophyte obtain its food (is it autotrophic or heterotrophic)?

How does the sporophyte obtain its food (is it autotrophic or heterotrophic)?

Which phase of the life cycle is larger and lives longer?

E X E R C I S E   1 2 . 4
Vascular Seed Plants—Gymnosperms

Objectives

After completing this exercise, you should be able to

1. Identify representative members of the four plant divisions of vascular seed plants.
2. Define homospory and heterospory and give examples of plants showing each condition.
3. Identify in living material or on a diagram all of the structures and events in the pine life cycle.
4. Identify and discuss the three important developments in plant evolution seen in the pine.

The remaining plant groups, the gymnosperms and angiosperms, share the evolutionary innovation of having seeds. In seedless plants, the minute, single-celled spores must leave the parent plant and find a suitable habitat to germinate and develop into gametophytes. In seed plants, the spores remain within the parent and the gametophytes that develop from the spores are produced within enclosed structures. Instead of the spore, it is the seed, equipped with a protective coat and a supply of food for the developing embryo, that ventures forth from the parent to establish a new plant. In gymnosperms the seeds are "naked" (Greek *gymnos,* naked; *sperma,* seed), while in angiosperms (Exercise 12.5) they are protected in containers (Greek *angion,* container).

When all of the spores produced by a plant are alike, the plant is said to be **homosporous** ("same spores"). If two kinds of spores are produced, the plant is **heterosporous** ("different spores"). Heterospory first evolved in seedless vascular plants, but most members of that group, including the representative life cycle you observed in lab, are homosporous. All of the seed plants are heterosporous. This makes their life cycles more complex, since there are "male" and "female" spores as well as male and female gametes.

The vascular seed plants also mark the appearance of **pollen,** which is an immature male gametophyte. Gymnosperms are free from the requirement of water for fertilization because the entire male gametophyte travels by air to the female gametophyte before the sperm are produced. Thus the sperm do not have to swim to the egg.

Use the margin of your lab manual to sketch the specimens on display in the lab.

Activity A: Cycads

These palmlike plants occur in subtropical and tropical regions; one species is common in Florida. An example is shown on Color Plate 2. Although now limited to 100 species in 10 genera, they were quite numerous in the Mesozoic era, 245–65 million years ago, when they served as dinosaur food. More recently, their underground stems and root-stalks were eaten by the Seminole Indians.

Activity B: Ginkgo

Ginkgo biloba (the maidenhair tree, so called because the leaves resemble maidenhair fern) is the only remaining species of this once common group. Unlike most gymnosperms, it is deciduous; the leaves, illustrated on Color Plate 2, are all shed in autumn. This tree has been cultivated for centuries by the Chinese and Japanese and is now widely planted in the United States, where its general hardiness and tolerance of air pollution make it a suitable city dweller. Like many trees, a ginkgo has either male *or* female reproductive parts. Male trees are preferred (and easily obtained by vegetative propagation) because the fleshy seed coat produced by the females has an odor similar to rancid butter.

Activity C: Conifers

The largest group of gymnosperms is the conifers or cone-bearers. Included in this group of 550 species are such economically important plants as pines, firs, and spruces. Conifer leaves have reduced surface area, sunken stomates, and thick cuticles, features that help resist drought.

You will look at a typical pine life cycle. This is the first heterosporous life cycle you have encountered, and you will immediately notice that it is more complex than the homosporous moss and fern life cycles. In addition to the male and female gametes that are produced by the male and female gametophytes, heterosporous plants also produce male and female spores. The key to understanding heterosporous life cycles is to follow the male path from beginning to end and then go back and start over with the female path.

All of the events of the conifer alternation of generations life cycle take place in reproductive structures called cones. The sporophyte produces two kinds of cones, male and female. The spores that are produced by meiosis in sporangia in male cones undergo two mitotic divisions to become the male gametophytes, which are also known as pollen grains. In certain areas of the country it is easy to appreciate just how prodigiously pines produce pollen: In the spring a thick yellowish coating is apparent on cars and sidewalks. The pollen is carried by the wind to the female cones of other trees; the female cones exude a sticky resin to trap it.

The scales of the female cones bear ovules, which enclose the female sporangia. Cells in the female sporangia also undergo meiosis to produce spores. One of the spores from each meiotic division develops into a female gametophyte by mitosis. As the pollen germinates and produces sperm nuclei by mitosis, the female gametophyte produces an egg. One sperm nucleus reaches the egg through the pollen tube and fertilizes it. Notice in Figure 12.6 that the male gametangium has been eliminated, and the sperm is now merely a nucleus rather than a motile cell.

Through mitotic divisions, the zygote becomes an embryo. The female gametophyte remains as a food supply for the developing embryo to use when the seed germinates, and a protective seed coat is formed by the surrounding sporophyte tissues in the pine cone.

Procedure

1. Look at Figure 12.6 and label the arrows that indicate where mitosis, meiosis, and fertilization occur.
2. Identify the following structures in the living material and label them on Figure 12.6:

Sporophyte	Male gametophyte (pollen)
Female spores	Sperm nucleus
Female gametophyte	Zygote
Egg	Embryo
Male spores	

3. With a colored pencil, color all of the haploid structures in Figure 12.6.

   Describe the difference between pollination and fertilization.

How does the sperm get to the egg?

How does the gametophyte obtain its food (is it autotrophic or heterotrophic)?

How does the sporophyte obtain its food (is it autotrophic or heterotrophic)?

Which phase of the life cycle is larger and lives longer?

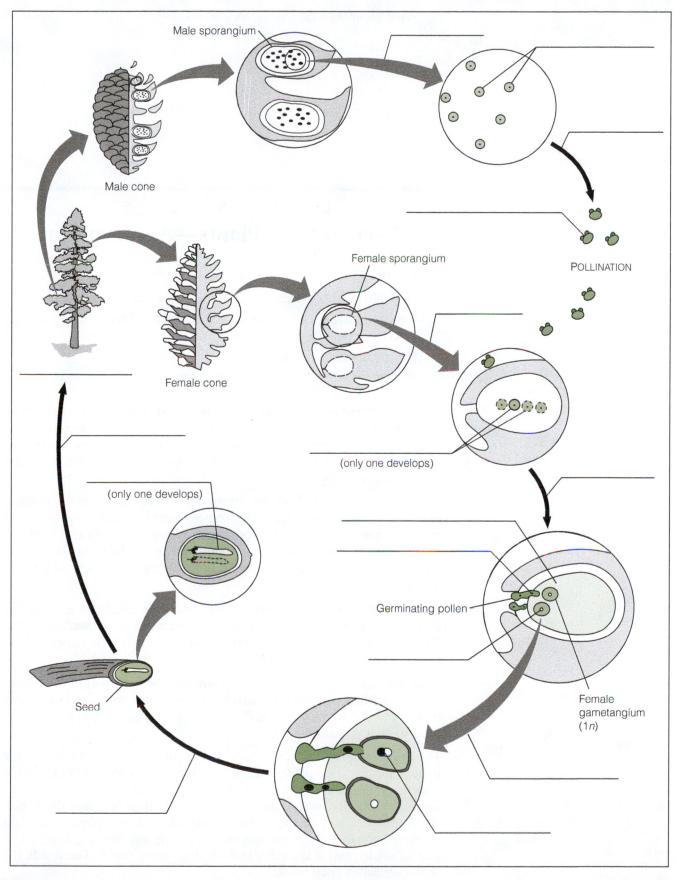

Male sporangium

Male cone

POLLINATION

Female sporangium

Female cone

(only one develops)

(only one develops)

Germinating pollen

Female gametangium (1n)

Seed

Figure 12.6.
Pine life cycle.

What advantages do seeds have over spores?

E X E R C I S E   1 2 . 5
Vascular Seed Plants—Angiosperms

Objectives

After completing this exercise, you should be able to

1. Identify in living material or on a diagram all of the structures in the flowering plant life cycle.
2. Explain double fertilization.

The key characteristic of angiosperms is the flower. Any plant that produces flowers is a member of this group, which is also known as the flowering plants. The name angiosperm, which means "seed container," refers to the fact that the female gametophytes and then the seeds are enclosed in an ovary.

The diversity seen in all of the previously described groups is nowhere near that of the angiosperms. They range in size from the aquatic water meal at 1 mm to the giant eucalyptus trees. Flowering plants are found in a diverse array of habitats. Of the 280,000 species of plants, approximately 250,000 are angiosperms, making them by far the most successful group of plants.

Just as all the events of the conifer life cycle take place in the cone, the flower houses the entire alternation of generations for flowering plants. The flower, which is produced by the sporophyte, may have either male or female reproductive parts or both.

Male sporangia are located within the **anther,** the male part of the flower. After the male spores are produced by meiosis, each spore undergoes one mitotic division to produce a pollen grain, or male gametophyte. As in the pine, there is no gametangium.

The female reproductive structure is the **ovary.** Inside the ovary are **ovules.** Initially the ovules are the female sporangia. One cell in each ovule undergoes meiosis to produce spores. One of the spores enlarges and divides mitotically several times, and the ovule then becomes the female gametophyte. There is no gametangium. The gametophyte has eight nuclei. At one end of the gametophyte is the egg nucleus, flanked by two smaller nuclei, and at the other end are three more nuclei. The two in the center are called polar nuclei.

Pollen, which is the male gametophyte, is brought by wind, water, or animal to the female part of another flower. The pollen then germinates, and a long pollen tube containing two sperm nuclei grows down to the ovule. One of the sperm nuclei fuses with the egg nucleus to form a diploid zygote. The other sperm nucleus fuses with the two polar nuclei to form an endosperm nucleus, which is triploid because it is the fusion of three haploid nuclei. Because of these two separate fertilization events, flowering plants are said to have **double fertilization.**

The endosperm nucleus undergoes many mitotic divisions to produce a multicellular **endosperm,** which serves as a food source for the developing plant. The zygote divides and differentiates to form an embryo, and the layers of cells that surround the ovule harden into a protective seed coat. The ovary itself later matures into a fruit, which helps protect and disperse the seeds.

Procedure

1. Observe the specimens on display that represent the angiosperm life cycle and match them with Figure 12.7 on the next page.

2. On the diagram, label the arrows that indicate where meiosis, mitosis, and fertilization occur.

3. Identify the following structures:

Sporophyte	Egg
Male spore	Sperm nuclei
Male gametophyte	Embryo
Female spore	Zygote
Female gametophyte	

4. With a colored pencil, color all of the haploid structures in Figure 12.7.

How does the sperm get to the egg?

How does the gametophyte obtain its food (is it autotrophic or heterotrophic)?

How does the sporophyte obtain its food (is it autotrophic or heterotrophic)?

Which phase of the life cycle is larger and lives longer?

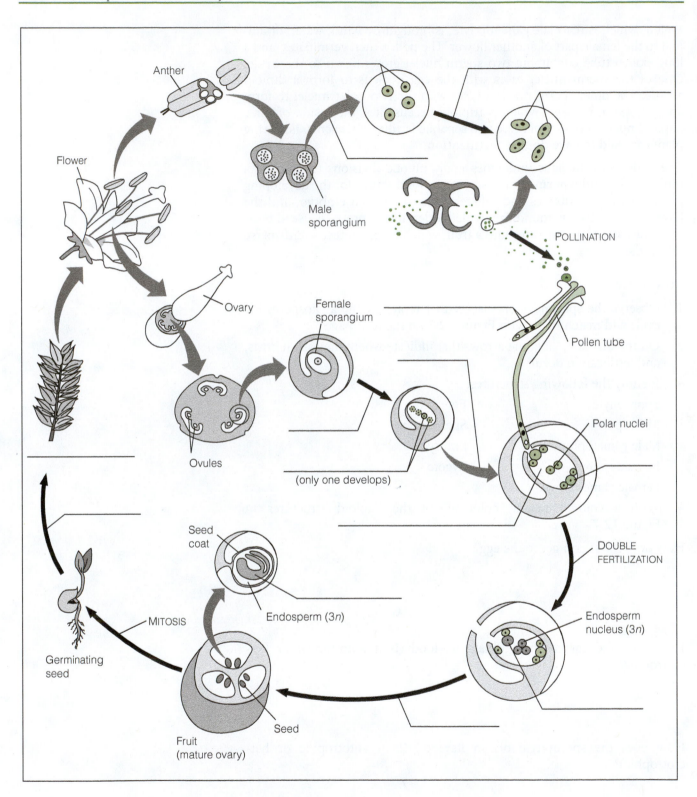

Figure 12.7.
Angiosperm life cycle.

What is the advantage of having seeds produced in an ovary?

Review the diagrams of the moss, fern, pine, and flowering plant life cycles on which you have colored the haploid structures. What can you say about changes in the predominance of the haploid and diploid forms in the course of evolution?

Questions for Review

1. Why can only diploid cells undergo meiosis?

2. Why are the bryophytes dependent on water for reproduction?

3. If a moss gametophyte has a sporophyte attached to it, is the gameto-phyte male or female? How do you know?

4. What characteristics would you use to distinguish a fern from a horsetail?

5. How are club mosses different from true mosses?

6. Are ferns as dependent on water as mosses are? Explain.

7.   Seedless vascular plants related to the club mosses and horsetails were once dominant forest components. When the world's climate became cooler and drier, the seedless plants were eclipsed by seed plants. Why would the seed plants have been more successful in the new climatic conditions?

8.   How did gymnosperms achieve independence from the need for water for fertilization?

9.   What is the advantage of an ovary?

10.   Compare the way seeds are made in gymnosperms with the way seeds are made in angiosperms.

Animal Diversity

Introduction

Our study of the animal kingdom in this lab topic will include nine of the major phyla, shown in Figure 13.1 on the next page. There are 26 other phyla (depending on which taxonomist you ask) that have been omitted. When you recall the classification hierarchy by which similar organisms are grouped together (domain, kingdom, phylum, class, order, family, genus, species), you will realize that we are painting the animal kingdom in very broad strokes. We are also able to sample only a bit of the diversity—this is a kingdom that reaches all the way from the barely organized sponges through that most complex of animals, the human.

The characteristics that distinguish animals from other organisms include multicellularity and the absence of cell walls. Most animals have the ability to move and some type of nervous system. Animals are also distinguished by their means of sexual reproduction and their life cycles. The animal characteristic that will be explored in more depth for the phyla studied in this lab topic is the heterotrophic mode of nutrition. That is, animals derive their nutrients from feeding on others (*hetero* = other; *trophic* = feeding).

Almost all animals take food into their bodies and then digest it internally. The remaining few are parasites that obtain nutrition from food that has already been digested by their hosts. You probably already know that animals can be carnivores (meat-eaters), herbivores (plant-eaters) or omnivores (eat both plants and animals). But there are also some animals that feed on decaying organic matter. And a large number of aquatic animals are suspension feeders (sometimes called filter feeders) that pass currents of water through their bodies and trap food particles from it.

Outline

Exercise 13.1: Phylum Porifera—Sponges

Exercise 13.2: Phylum Cnidaria—Jellyfish, Sea Anemones, Corals

Exercise 13.3: Phylum Platyhelminthes—Flatworms

Exercise 13.4: Phylum Nematoda—Roundworms

Exercise 13.5: Phylum Annelida—Segmented Worms

Exercise 13.6: Phylum Mollusca—Mollusks

Exercise 13.7: Phylum Arthropoda

Exercise 13.8: Phylum Echinodermata

Exercise 13.9: Phylum Chordata

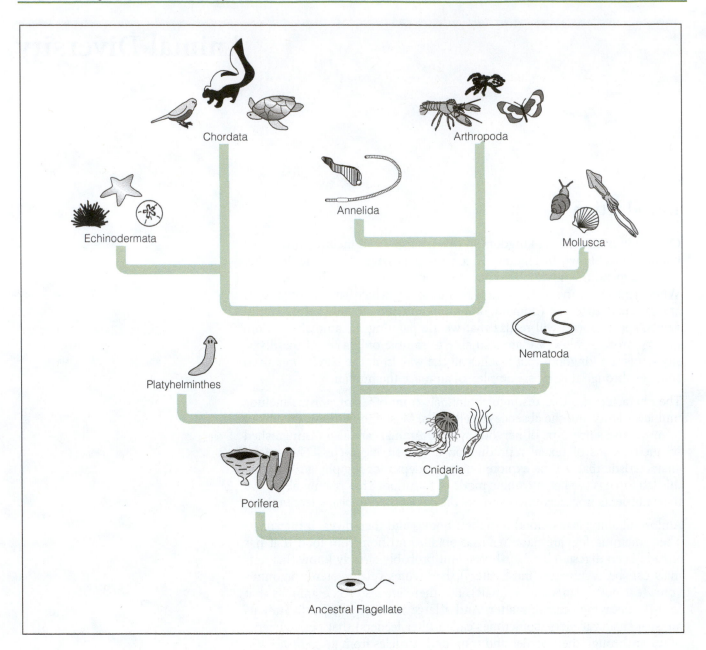

Figure 13.1.
Phylogenetic tree showing 9 of the 35 animal phyla.

EXERCISE 13.1
Phylum Porifera—Sponges

Objective

After completing this exercise, you should be able to
1. Describe how sponges obtain food.

The phylum Porifera (Latin *porus*, pore; Greek *fera*, bearing) comprises the sponges. As you can see in Figure 13.1, the sponges diverged early from the main line of animal evolution and have given rise to no other animal groups. There are approximately 9,000 species of sponges. All sponges are aquatic and most of them are marine. A vividly colored sponge is shown on Color Plate 3; its pigmentation is due to symbiotic algae. The adult forms are sessile, which means that they are stationary, but the larvae (immature animals that do not resemble the adults) are motile.

Figure 13.2 shows a diagram of a typical sponge. The body is composed of two cell layers: an outer epidermal layer and an inner layer of flagellated "collar cells." Between the epidermal layer and the layer of collar cells are amoebalike cells (amoebocytes) with various functions such as food storage, digestion, and formation of reproductive cells. Some of the amoebocytes secrete materials that form an endoskeleton, which supports the sponge. One skeletal material is a protein that forms a network of tough fibers. The other type of skeletal material is mineral crystals called spicules. Some sponges have both protein fibers and spicules and some have one or the other. For example, bath sponges are fibrous endoskeletons that have no spicules. (Luffa sponges are not animals at all, but a plant related to squash.)

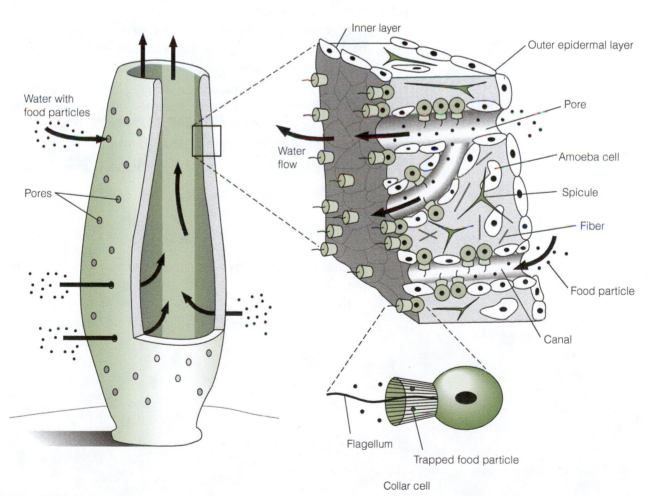

Figure 13.2.
Anatomy of typical sponge.

Sponges are suspension feeders. Their bodies are perforated with the pores for which the phylum is named, and water flows into the sponge through them (Color Plate 3). There can be a remarkable number of pores. It is estimated that *Leuconia*, a sponge measuring about 10 cm tall and 1 cm in diameter, has about 81,000 of them! The water travels through a network of canals lined by collar cells (Figure 13.2) whose beating flagellae bring the water in. The "collar" of collar cells is a molecular mesh that sieves out food particles such as microscopic algae, bacteria, and organic debris. The filtered water then proceeds into the central chamber where it is expelled through a large opening called an osculum.

In the following procedure, you will observe the structure of a small marine sponge called *Grantia* (also known as *Scypha*).

Procedure

1. Get a preserved specimen of *Grantia* from your instructor and place it in a small petri dish.
2. Observe the specimen under the dissecting microscope. First use reflected light, then transmitted light.
3. Sketch the sponge in the margin of your lab manual and label the osculum.
4. Use a razor blade or scalpel to slice the sponge longitudinally so that you can see the central canal.
5. Place the specimen with the central canal facing up and look at it under the dissecting scope, again trying both light sources.
6. Sketch the specimen in the margin of your lab manual.
7. Use a compound light microscope to examine a prepared slide of *Grantia*. This thin cross section has been stained so you can see the collar cells protruding into the canal. You may also be able to see spicules.
8. Sketch the specimen in the margin of your lab manual.

Given that all sponges are filter feeders, why does it follow that all sponges are aquatic?

Would mobility improve the ability of sponges to capture food? Explain.

EXERCISE 13.2
Phylum Cnidaria—Jellyfish, Sea Anemones, Corals

Objectives

After completing this exercise, you should be able to

1. Name representative members of the phylum Cnidaria.
2. Sketch the polyp and medusa body forms.
3. Describe the feeding behavior of cnidarians in general and hydra in particular.

There are approximately 10,000 species in the phylum Cnidaria (Greek *knide*, nettle; Latin *aria*, like). They are all aquatic (most of them are marine), and they are all radially symmetrical. An object or an animal that is radially symmetrical can be equally divided in more than one way because its parts are arranged around an imaginary central axis. The hydra, a cnidarian shown in Figure 13.3, is an example. It can be divided into equal halves starting at any point on the perimeter, as long as the dividing line passes through its center.

Figure 13.3.
Radial symmetry.

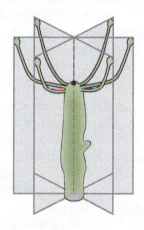

Cnidarians have two distinct body forms: the polyp (for example, sea anemones), which is stationary, and the free-floating medusa (for example, jellyfish). Figure 13.4 shows a comparison of these forms. There are two layers of cells separated by a gelatinous material that helps support the body. Note the increased amount of this material in the medusa form—this is the "jelly" of jellyfish.

Some cnidarians have both body forms in their life cycle. If an animal is adapted to a stationary way of life, what could be the purpose of having a motile phase at some point in the life cycle?

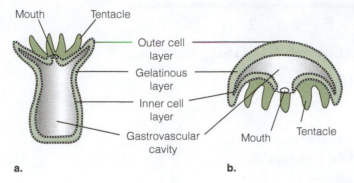

Figure 13.4.
Cnidarian body forms: (a) polyp form; (b) medusa form.

Sea anemones and corals are polyp forms of Cnidaria. Color Plate 3 shows a sea anemone waving its tentacles. The anemone's body is filled with water to provide support by hydrostatic pressure. The actual mass of tissue comprising the body is quite small. When the animal is endangered, much of the water is expelled and the anemone shrinks remarkably, drawing in its tentacles. The coral polyp itself looks very similar to an anemone. By precipitating minerals from seawater, some types of coral encase themselves in a hardened exoskeleton (see Color Plate 3). Colonial corals have left behind vast stretches of coral reefs in warm, shallow waters throughout the world. The Great Barrier Reef extends for 1,800 km along the coast of northeastern Australia and is the largest structure ever built by living organisms.

Most species of jellyfish spend most of their lives in the medusa form and the polyp stage is reduced or absent. Although most jellyfish range from 2 to 40 cm in diameter, the largest is over 2 m across and has tentacles 60 to 70 m long. Movement of jellyfish is by a beautiful rhythmic undulation of the bell, or umbrella, of the medusa, with the tentacles trailing languidly beneath. A photograph of a jellyfish appears on Color Plate 3.

Cnidarians are carnivorous, consuming prey ranging from minute zooplankton to fish. They are armed with stinging cells called **cnidocytes**, which are used in prey capture and in defense. These stinging cells are made up of coiled tubular threads, many with a barb or spine. When it receives certain chemical or tactile stimuli, the thread springs out, harpooning or entangling the prey. Some can inject poison as well. It is the stinging cells of the Portuguese man-of-war and certain jellyfish that make contact with these animals painful and sometimes dangerous for humans. The phylum name Cnidaria refers to a plant (nettle) with stinging cells that the plant uses for defense.

The hydra is a freshwater polyp form that is a popular laboratory organism. Hydra feed on insect larvae, small crustaceans, and annelid worms.

Procedure

1. Use a dropper to get a hydra from the container marked "hungry." Put it into a small watch glass with several drops of water from the hydra's container. (You may have to flush the hydra out of the dropper by sucking water into the dropper and then rapidly squirting it out.)

2. Examine the hydra under the dissecting microscope and sketch it in the margin of your lab manual. Reflected light works best.

3. Dip a piece of cotton thread into the dish and touch the hydra's tentacles with it. Describe what happens.

4. Soak the thread in liver juice and dip it into the water near the hydra. Describe what happens.

What type of sensory information enables the hydra to locate its prey?

5. Get a brine shrimp or *Daphnia* (water flea) and put it in the dish with the hydra. Describe the hydra's feeding response.

✳ **When you have finished with the hydra, put it in the container marked "fed," whether it consumed the prey item or not.**

The digestive system of cnidarians has only one opening, which leads into the gastrovascular cavity. The undigested remains of the hydra's meal will later be ejected from the same opening that took the food into the body.

Why is this type of digestive system inefficient?

Why is radial symmetry adaptive for a sessile animal in an aquatic habitat?

EXERCISE 13.3
Phylum Platyhelminthes—Flatworms

Objectives

After completing this exercise, you should be able to

1. Name representative members of the phylum Platyhelminthes.
2. Explain why bilateral symmetry is more advantageous for a motile animal than radial symmetry.
3. Describe the feeding behavior of the planarian.
4. Explain how a tapeworm obtains food.

The phylum Platyhelminthes (Greek *platys,* flat; Greek *helmins,* worm) includes about 20,000 species. One unifying characteristic is the shape of these animals: Their bodies are flattened. Although they have a primitive one-opening digestive system like that of the cnidarians, this phylum has some features that are important in more advanced animals. For example, there are organs (groups of tissues that work together to perform a function) and division of labor among the organs.

Free-Living Flatworms

For this lab activity, you will observe the planarian, a freshwater flatworm. Although it is small and drably colored, its close marine relatives can be much larger (up to 30 cm), with vivid coloration. An example of a marine flatworm is shown in Color Plate 4.

Procedure

1. Get a planarian from the "hungry" container and observe it under the dissecting microscope. Make a sketch of it in the margin.
2. With only the reflected light source on, cover the side of the dish opposite the planarian with a piece of paper. What does the planarian do?

3. Put a tiny bit (just as much as you can pick up with the tips of the forceps) of egg yolk or liver in the dish with the planarian. Turn off the light while you wait for the animal to begin feeding. Describe its behavior.

✳   **When you have finished with the planarian, put it in the "fed" container whether it has eaten or not.**

What can you conclude about the sensory capabilities of the planarian?

Bilateral symmetry is seen in the adult form for the first time in the flatworms. In a bilaterally symmetrical animal there is only one way in which to divide the animal so that both halves are the same size and shape. Humans and crayfish (Figure 13.5) are other examples of bilaterally symmetrical animals.

Figure 13.5.
Bilateral symmetry.

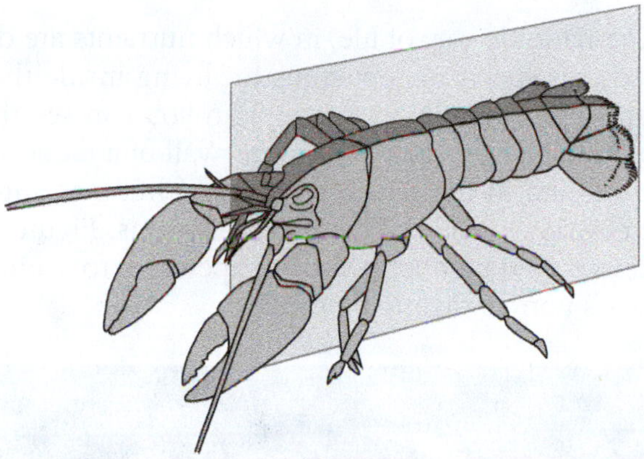

After observing the planarian's behavior, why is bilateral symmetry more adaptive than radial symmetry for a motile animal? (Hint: How is the body form of the planarian different from that of the hydra?)

Parasitic Flatworms

Free-living flatworms such as planaria are actually a minority, comprising less than a quarter of all platyhelminth species. The rest are parasitic. Liver, lung, and blood flukes are examples of parasitic flatworms. A type of blood fluke causes schistosomiasis, a disease that is a serious threat to public health in parts of much of Africa as well as elsewhere (Figure 13.6). Swimmer's itch is caused by a related species that lives in northern North American lakes.

Figure 13.6.
Scanning electron micrograph of *Schistosoma mansoni.*

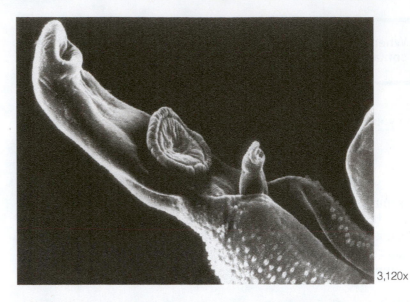

3,120x

The parasitic way of life, in which nutrients are derived from a living host, requires specific adaptations for living inside the body of another organism. For example, in Figure 13.6 you can see the sucker with which the blood fluke attaches itself to the wall of a blood vessel. Tapeworms, which are found in the intestinal tracts of humans and many other vertebrates, are another group of parasitic flatworms. Figure 13.7 shows the grappling device used by one species of tapeworm to avoid being swept away by the host's normal digestive processes.

Figure 13.7.
Scanning electron micrograph of the head end of a tapeworm showing hooks and suckers for attachment.

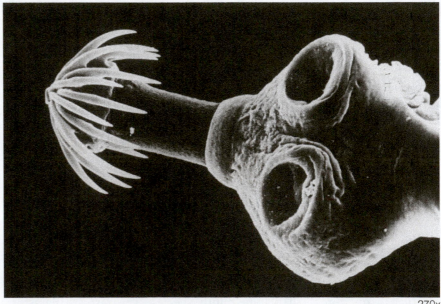

270x

Procedure

Observe the preserved tapeworm specimen on display. Sketch it in the margin of your lab manual and describe it below.

Reproduction can be an extremely complex matter for internal parasites. Why is this so? (Hint: What's the first requirement for sexual reproduction?)

On the other hand, methods of obtaining and digesting food may be extremely simplified. In fact, internal parasites may have no digestive system at all. How can a tapeworm, for example, survive without a digestive system? (Hint: What is the purpose of digestion?)

EXERCISE 13.4
Phylum Nematoda—Roundworms

Objectives

After completing this exercise, you should be able to

1. Name representative members of the phylum Nematoda.
2. Explain the advantage of having a complete digestive tract.

Although only 90,000 species of nematodes (Greek *nema*, thread; Greek *eidos*, form) have been identified, it has been estimated that there are close to half a million, living in all habitats and following various lifestyles. Where they occur (and they occur almost everywhere), their numbers are staggering. The top six inches of soil in one acre of a plowed field is estimated to contain billions of nematodes.

Nematodes are bilaterally symmetrical. They are generally small (less than 1 cm long). Free-living nematodes eat other nematodes or microorganisms, or they may feed on decaying organic material. Although most species are free-living, some are parasitic on animals (including humans) and plants. The parasitic species have been more thoroughly studied than the free-living nematodes. Examples of parasitic nematodes are hookworms, pinworms, *Ascaris* (intestinal roundworms), and filarial worms (including those known as heartworms in dogs). Some parasitic nematodes may exceed 8 cm in length.

Procedure

Observe the nematodes on display and note the characteristics they share.

The nematodes are the first animals we have seen in this laboratory that have a complete digestive tract—that is, one with two openings, a mouth and an anus (Figure 13.8). The hydra and planaria have an incomplete digestive tract.

Figure 13.8.
Complete digestive tract with little specialization.

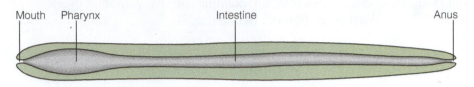

Mouth Pharynx   Intestine   Anus

What is the advantage of a two-opening digestive tract?

EXERCISE 13.5
Phylum Annelida—Segmented Worms

Objectives

After completing this exercise, you should be able to

1. Name representative members of the phylum Annelida.
2. Describe the characteristics that distinguish annelids from the flatworms and roundworms.
3. Describe the earthworm's digestive system.

There are about 15,000 species of annelids (Latin *annellus*, little ring), including marine, freshwater, and terrestrial members. Both free-living and parasitic lifestyles are represented. The earthworm is the most familiar example of an annelid, and it illustrates why this group is called the segmented worms. The annelid body consists of a head, a segmented body, and a terminal portion. The body segments are partitioned off from each other and are filled with a fluid that serves as a hydrostatic skeleton.

The largest group of annelids is the **polychaetes.** Most of them are marine and they are an important source of food for fish and crustaceans. Polychaetes have a pair of fleshy appendages on each body segment (Figure 13.9). Note that we did not see any type of appendages (external organs) in any of the more primitive phyla. In more highly organized animals, for example the Arthropoda, we will see greater specialization of these structures. The polychaete shown in Figure 13.9 is a clam worm. These animals hide out in burrows during the day and venture forth at night to search for

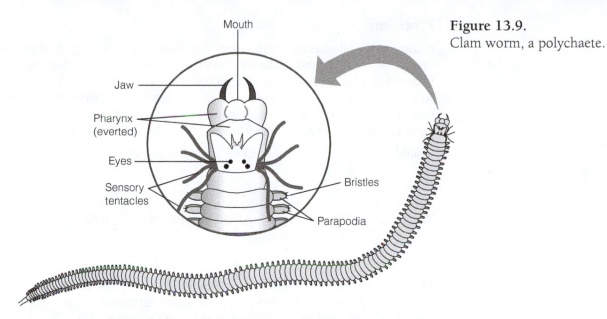

Figure 13.9.
Clam worm, a polychaete.

small animals (for example, other types of worms) to eat. Notice the predatory jaws. Other polychaetes, including the feather duster worms, build tubes to live in. They use tentacles covered with cilia to trap food such as tiny animals and decaying organic matter and transport it to the mouth.

Leeches comprise another group of annelids. Mostly found in fresh water, leeches include carnivorous species as well as the famous ecto- (external) parasites that feed on blood.

A third group of annelids includes the **earthworm.** These familiar animals feed on decaying organic matter in the soil and play an important ecological role as decomposers. Earthworms prefer a moist, humus-rich soil that they suck in through the pharynx as they burrow. Nutrients are extracted from it by the digestive tract. In marked contrast to the relatively unspecialized nematode digestive system, the earthworm digestive tract is divided into several different organs (Figure 13.10).

Figure 13.10.
Earthworm.

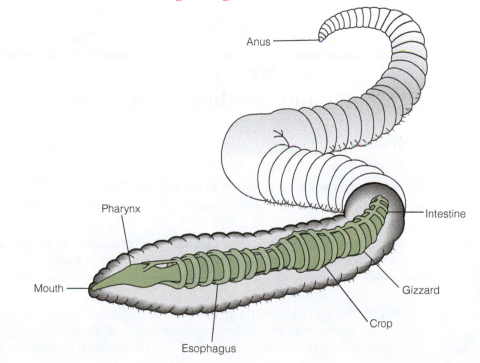

Procedure

Examine the dissected earthworm and note its specialized digestive organs. At the front end of the earthworm are the mouth and pharynx. The pharynx is a muscular organ used to draw food into the mouth. The next structure is the relatively long esophagus, a passageway between the pharynx and the crop, where food is stored temporarily. From the crop, food passes on to the gizzard where it is ground into smaller pieces.

Would you expect the gizzard to be muscular or nonmuscular? Why?

What is the advantage of having food broken up into smaller pieces?

The next part of the earthworm's digestive system is the intestine, which runs the remainder of the length of the worm. Enzymatic digestion takes place here, followed by absorption of the nutrient molecules.

Why is the intestine the longest organ in the digestive system?

Finally, undigested material is expelled through the anus.

How does specialization of organs in the digestive system increase the efficiency of digestion?

EXERCISE 13.6
Phylum Mollusca—Mollusks

Objectives

After completing this exercise, you should be able to

1. Name representative members of the phylum Mollusca.
2. Explain how you would distinguish among gastropods, bivalves, and cephalopods.
3. On a dissected specimen or a diagram of a clam, identify the mantle, gills, foot, and visceral mass and trace the pathway of food.

With 150,000 species identified, the phylum Mollusca (Latin *molluscus*, soft) is second only to the Arthropoda in number of species. Among the mollusks we find animals ranging in size from microscopic to gigantic, and we find fast-swimming predators, passive suspension feeders, peaceful herbivores, and parasites. Although most mollusks are marine, the phylum does include some freshwater species, and one group (snails and slugs) has been successful on land. This phylum also includes a number of species that humans enjoy as food: clams, mussels, oysters, scallops, squid, and snails.

The basic mollusk body plan includes a foot, which is mainly muscular tissue, and a visceral mass where the digestive, circulatory, respiratory, and reproductive organs are located. The mantle consists of two folds of skin that sandwich the visceral mass. The cavity created in between serves important functions in each molluscan group. In many species, the outside of the mantle secretes a protective exoskeleton, the shell. Gills extract oxygen from the water. In suspension feeders, they also collect food particles. Figure 13.11 shows the variations of body plan seen in the three major classes of mollusks.

Phylum Mollusca is categorized into classes based on the shape and use of the foot. One class is the **gastropods.** "Gastropod" means "belly foot." This group, the most abundant and diverse group of mollusks, includes snails, periwinkles, abalones, slugs, limpets, and others. Most gastropods have a single shell, for example the marine snail shown in Color Plate 4. Some gastropods, however, such as the slugs and sea hares, lack shells altogether (see Color Plate 4). The majority of gastropods are herbivores and possess a radula, which is a set of rasping teeth arranged in a tonguelike organ (see Figure 13.12). The radula is unique to mollusks. Watch a snail as it grazes on the side of an aquarium. It is scraping algae off the glass with its radula.

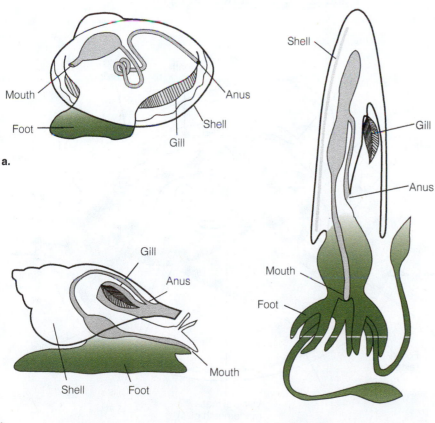

Figure 13.11.
Molluscan body plans: (a) clam; (b) snail; (c) squid.

a.

b.

c.

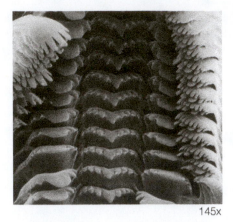

145x

Figure 13.12.
Scanning electron micrograph of a snail radula.

Carnivorous gastropods include the oyster borer, which uses its radula to drill through the oyster's shell, and snails of the genus *Conus* (Color Plate 4), which have a radula that is specialized for stinging prey. The venom from some *Conus* species is even fatal to humans.

The squids, octopuses, and nautiluses are **cephalopods**, which means "head foot." All of these animals have well-developed heads with very advanced eyes. The foot is modified into tentacles with suction cups for grasping prey. Color Plate 4 shows a squid. Cephalopods are all carnivores and are found exclusively in marine habitats. Except for the nautilus, the shells of cephalopods are much reduced. In squids and cuttlefish the shell is small and enclosed in the mantle (the internal shell of the cuttlefish is the cuttlebone sold as a beak-sharpening device to be hung in bird cages), and the octopus has no shell at all.

Members of the **bivalve** class are "hatchet-foot" mollusks that use the foot for burrowing. Bivalves have two shells. Some familiar examples are clams, scallops, and oysters. Bivalves lack a head, eyes (except in scallops), and radula. Most of them are suspension feeders. Water enters through an opening called a siphon, and passes over the gills, which collect food particles from the water. The water is then expelled through a second siphon. Color Plate 4 shows a clam with its foot and both siphons extended outside the shell.

Procedure

Examine the dissected bivalve specimen on display and identify the mantle, visceral mass, gills, and foot (see Figure 13.13).

Next, trace the path of food through the clam. Locate the structures on the dissected clam and sketch the pathway on Figure 13.13.

Figure 13.13.
Dissected clam. Top half of mantle has been cut away to reveal foot and gills.

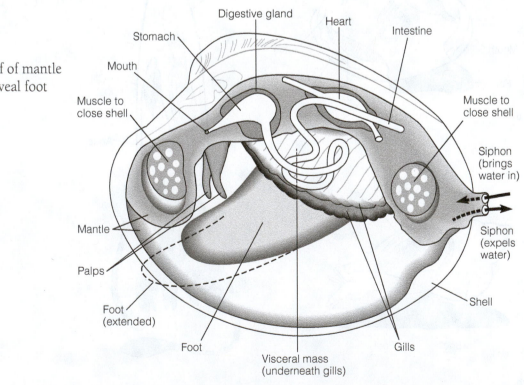

Water enters through the lower siphon and is drawn to the gills by the beating of cilia. Food particles are sieved out by the gills and conveyed by mucus threads to the palps, which direct the food to the mouth. Once in the mouth, food passes through the digestive system, including the esophagus, stomach, intestine, and anus. Meanwhile, the filtered water passes out through the other siphon.

Compare the clam's digestive system to that of the earthworm. Why do earthworms have two organs that clams don't have?

E X E R C I S E   1 3 . 7
Phylum Arthropoda

Objectives

After completing this exercise, you should be able to

1. Identify representative members of the five groups of arthropods.
2. Describe the structural features that are characteristic of each of the five groups of arthropods.

With 1.1 million species, arthropods (Greek *arthron*, joint; Greek *pous, podos*, foot) are the most numerous of all animals. In fact, 85% of all animals are arthropods, 75% of all animals are insects (a class of arthropods), and 30% of all animals are beetles (an order of insects). Arthropods are also the most diverse group of animals. They are widely distributed and fill an array of ecological niches.

The name of the phylum Arthropoda refers to the jointed appendages that are characteristic of members of this group. Some of the appendages are used for locomotion, but in many groups appendages have been modified for food handling, defense, taste, touch, and reproduction. All of the arthropods are bilaterally symmetrical. Like the annelids, their bodies consist of a series of segments.

Arthropods have a chitinous exoskeleton, which is one factor that allowed arthropods to be successful in terrestrial habitats. Chitin is a tough but flexible polysaccharide that protects the soft body parts and prevents dehydration. It may be thick and armorlike (as in beetles) or soft (as in spiders). Even a hardened skeleton is soft at joints and between body segments so the body is flexible. Since the exoskeleton cannot grow, arthropods periodically shed it and grow a larger one, a process called molting.

The major groups of this diverse phylum are briefly described in this lab exercise.

The **crustaceans** include the lobster, crayfish, shrimp, *Daphnia* (water fleas), barnacles, crabs, and pill bugs (see Color Plate 5). Most crustaceans are aquatic; only the pill bug is terrestrial.

The exoskeleton of most crustaceans is supplemented by calcium salts. This is seen to an extreme degree in lobsters and crabs. Consequently, their shells are heavier and less flexible than those of other arthropods.

Thinking about various crustaceans you have encountered, how do you suppose this class got the name Crustacea?

Crustaceans provide an excellent illustration of the jointed appendages that gave arthropods their name.

Procedure

1. Use a dissecting microscope to examine the crayfish.
2. Sketch the crayfish in the space below.

3. Describe the exoskeleton of the crayfish.

4. Describe the jointed appendages of the crayfish.

The **arachnids** include spiders, scorpions, mites, and ticks (see Color Plate 5). Most arachnids are carnivorous and have claws, fangs, poison glands, or stingers. Since spiders have no jaws and are unable to chew, they inject their prey with digestive juices. Their mouthparts are adapted for drawing in the digested fluid.

Not all spiders capture their prey with webs. For example, there are the trap-door spiders, which dig a web-lined hole in the ground with a hinged door. The door serves both as protection from enemies and as a means of locating prey, which produce vibrations as they walk over the door. Other spiders, such as jumping spiders, wolf spiders, tarantulas, and fishing spiders, hunt for prey.

Insects are the most diverse and abundant class of organisms and can be found in many habitats. They exhibit all types of feeding behaviors. Though most are herbivorous, some are parasitic, some are carnivorous, and some feed on decaying organic matter. The mouthparts of insects are specially adapted for each insect's feeding habits, so they vary greatly from one species to another.

Humans have many unpleasant interactions with insects, including those that sting, bite, suck our blood, and transmit disease. Then there are the pests that infest our homes and gardens. You are about to examine an insect that causes a lot of damage, so to balance our coverage, can you name some insects that are beneficial to humans?

Termites are the damaging insects mentioned above, and they have an unusual digestive system. Termites feed on plant materials, including wood. But they lack the digestive enzyme needed to dismantle cellulose molecules, which are a major constituent of wood. The cellulose is digested by protists, single-celled eukaryotes that live in the termite's digestive tract. These protists belong to the group called protozoans, or animal-like protists. They are further classified as flagellates because their means of locomotion is long, whiplike structures called flagella. These inhabitants of the termite gut do have the necessary enzyme for digesting cellulose, and termites are able to absorb some of the products for their own nutrition. This type of relationship, where two organisms live in close association with each other, is called symbiosis. Since both partners in this case benefit, the relationship is further classified as mutualism.

Procedure

1. Get a termite and place it on a microscope slide.
2. Hold the termite by using forceps to grasp its body in the middle.
3. Use a second pair of forceps to grasp the very tip of the termite's abdomen (its "tail end").
4. Pull gently. You should see the termite's gut separate from its body.
5. Remove everything from the slide except the gut.
6. Put a drop of 0.9% NaCl on the gut, then a drop of Proto-Slo. Use a tip of the forceps to mix it around.

Why should you use a NaCl solution instead of water?

7. Add a cover slip and tap it lightly with the eraser end of a pencil to spread out the tissue.
8. View your slide under a compound microscope.
9. Sketch some of the protozoans you see.

EXERCISE 13.8
Phylum Echinodermata

Objective

After completing this exercise, you should be able to

1. Identify representative members of the phylum Echinodermata and describe their unique adaptations.

The 6,000 species of echinoderms (Greek *echinos*, sea urchin or hedgehog; Greek *derma*, skin) are all marine and they are all radially symmetrical as adults. Examples include sea stars, brittle stars, sand dollars, sea urchins, sea cucumbers, and sea apples. Echinoderms possess many unique adaptations. One is locomotion by means of a water vascular system (vascular refers to any vessels that convey fluid). Water pressure in the system is varied, causing the "tube feet" to extend or retract (see Figure 13.14). Echinoderms have an endoskeleton that may have spines protruding through the skin. When you pick up the shell of the sand dollar on the beach, for example, sometimes spines are still attached to it. The holes where the tube feet extended through the shell can be seen in a delicate pattern on the shell.

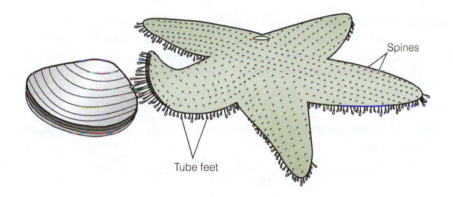

Figure 13.14.
Sea star.

Many sea stars are carnivorous, feeding on various invertebrates that either are sessile or move more slowly than they do. They are especially important predators of clams and oysters and can be an expensive nuisance to aquaculturists who farm these animals. Sea stars have an unusual way of feeding on bivalves (Color Plate 6). By wrapping its arms around the shell and pulling constantly with the suckers on its tube feet, the sea star is able to fatigue the muscles that are holding the shell closed just enough to create a small opening (less than 1 mm is sufficient). The sea star then turns its stomach inside out, inserts it into the shell, and digests the mollusk's soft body parts. When finished, it retracts its stomach into its own body.

The sea urchins, sand dollars, and sea biscuits have no arms; they are all spherical or disk-shaped. Their relationship with the star forms, however, can be seen if you picture the arms of a sea star bent backwards to meet at the tips (Figure 13.15). As in the other echinoderms, the body is enclosed in a shell that is covered by an epidermis. All of the members of this class are covered with movable spines (Color Plate 6). Although some species produce poison to paralyze prey, most sea urchins feed peacefully on algae or dead animals.

Figure 13.15.
Relationship of sea urchin (a) to sea star (b). Both sea stars and sea urchins have fivefold radial symmetry. Although sea urchins lack arms, there are five grooves where the tube feet are located.

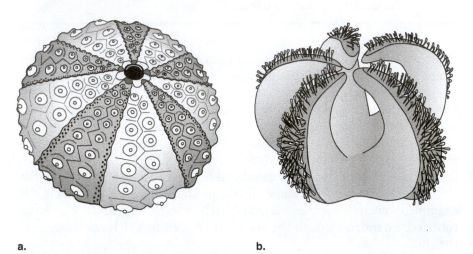

a.   b.

The sea cucumbers (Color Plate 6) and sea apples lack both spines and arms. The skeletal elements are small and are buried in the leathery body wall. These animals lie on the ocean floor and filter feed using branched tentacles that are really modified tube feet. Some sea cucumbers are burrowers.

Procedure

Observe the echinoderm specimens on display. Sketch them in the margin of your lab manual.

Why is radial symmetry adaptive for the echinoderms?

EXERCISE 13.9
Phylum Chordata

Objectives

After completing this exercise, you should be able to

1. Name representatives of the invertebrate members of the phylum Chordata.

2. Identify members of each of the five groups of vertebrates described in this exercise.

3. Describe the functions of the organs in the frog's digestive system.

4. Explain the significance of the amniotic egg.

5. Compare the digestive system of a typical bird to that of the earthworm and the frog.

6. Describe how the digestive tract of herbivorous mammals differs from that of carnivorous mammals.

The phylum Chordata includes a few invertebrates (animals without backbones) as well as the vertebrates. All share the common chordate characteristics: a hollow nerve cord that runs along the animal's back, a notochord, gill structures, and a post-anal tail.

One group of invertebrate chordates is the tunicates. Tunicates are marine animals commonly called sea squirts. Most of the adults are stationary suspension feeders. They are commonly found on docks, boats, and rocks, though some live in the open ocean. A group of tunicates is shown in Color Plate 6.

The largest group of chordates by far is the vertebrates, which are distinguished from invertebrates by the presence of a backbone. Although vertebrates make up only about 3% of all animals, they are probably the most familiar animals: fishes, amphibians, reptiles, birds, and mammals.

Fish are the most diverse group of vertebrates, including more than 30,000 species. The earliest fish had no jaws and survived as scavengers or parasites. The evolution of jaws brought unprecedented opportunities for exploiting food sources. The majority of fish are carnivorous. They seize and hold their prey with their teeth and then swallow it whole. Some have teeth in their throats to help grind the food. Herbivorous fish feed on aquatic plants such as grasses and other flowering plants and algae. Some fish eat both plants and animals, some are parasites, and some feed on organic debris. A large number of fish are suspension feeders. Many of them live in schools in the open sea, straining small floating plants and animals from the water.

Amphibians include approximately 4,800 species of newts, salamanders, toads, and frogs. As the name (Greek *amphi*, double; Greek *bios*, life) implies, amphibians lead a double life in terrestrial and aquatic habitats. Although important adaptations for terrestrial life have appeared, such as lungs and limbs, amphibians still require water—or at least moisture—for reproduction, and their larvae have an aquatic tadpole stage.

Most adult amphibians are carnivorous, feeding on invertebrates such as snails, worms, insects, and any other moving animal that is small enough to swallow. Frogs and toads have a tongue attached in the front of the mouth. When prey is spotted, the tongue is quickly launched and the prey is caught on the sticky end. The mouth has small teeth to hold the prey, which is then swallowed whole.

We will use the frog to illustrate the further specialization of the digestive tract in vertebrates. All vertebrates share similarities in their body plans, including in their internal organs, so the frog's digestive tract is very similar to our own.

Figure 13.16.
Digestive system of the frog.

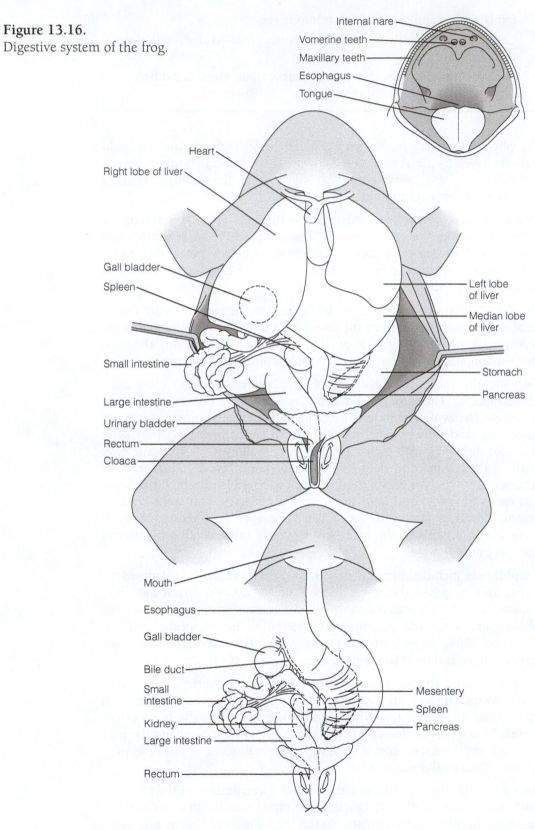

Internal nare
Vomerine teeth
Maxillary teeth
Esophagus
Tongue

Heart
Right lobe of liver
Gall bladder
Spleen
Left lobe of liver
Median lobe of liver
Small intestine
Large intestine
Stomach
Pancreas
Urinary bladder
Rectum
Cloaca

Mouth
Esophagus
Gall bladder
Bile duct
Small intestine
Mesentery
Spleen
Kidney
Pancreas
Large intestine
Rectum

Procedure

Examine the mouth of the frog. Note the teeth along the upper jaw and in the roof of the mouth. Does the frog use these teeth for chewing?

Examine the tongue of the frog and note where it is attached. The back of the mouth cavity—the throat—is the pharynx, which connects with the esophagus. Follow the path of the esophagus from the pharynx to the stomach. (If it has not already been done for you, remove most of the liver so you can see the esophagus.) Mixing and some enzymatic digestion of the food occur in the stomach.

Next locate the small intestine. Further enzymatic digestion occurs here, followed by absorption of nutrients into the blood capillaries for distribution to the body. The inner surface of the small intestine is very wrinkled, resulting in a large amount of surface area. Why is this important?

Find where the small intestine joins the large intestine. One of the main functions of the large intestine is to remove the water remaining in the material coming from the small intestine. From the large intestine, undigested remains go to the rectum to be expelled from the body.

Various glands are associated with the digestive system. The largest gland in the body is the liver. The liver produces bile, which emulsifies fats (Lab Topic 14). The bile is carried by a small duct to the gall bladder, where it is stored and released to the small intestine as needed. The gall bladder is beneath the liver and resembles a small green pea.

Also look for the pancreas, a thin, inconspicuous organ located in the membranous tissue between the stomach and small intestine. The pancreas provides the digestive enzymes needed by the small intestine.

Reptiles include approximately 6,500 species of snakes, lizards, turtles, and crocodiles. Reptiles were the first vertebrate group to achieve complete independence from the aquatic habitat. This was in part due to significant changes in reproduction. Amphibian eggs are encased in jelly and must be protected from drying out. Reptiles have a protective shelled egg that contains food and moisture for the developing embryo (Color Plate 6). This amniotic egg represents a major breakthrough for terrestrial life.

A related problem had to be solved at the same time. Most fish and amphibians release their gametes into the water, where sperm meets egg and fertilization occurs. Once the shell develops around the egg, which happens in the female's reproductive tract, there's no way for the sperm to get inside and fertilize it. The egg must be fertilized before it leaves the female's body. Thus all reptiles employ internal fertilization.

Anatomical differences also make reptiles better adapted to life on land. Their skeletons, for example, provide better support than do amphibian skeletons.

Why is structural support more important for a land-dwelling organism than it is for an aquatic one?

In contrast to the amphibians, reptiles have dry, scaly skin that protects against dehydration. Since the skin is no longer part of the respiratory system as it is in amphibians, the lungs are more efficient. The circulatory, excretory, and nervous systems have also undergone evolutionary changes to adapt reptiles to terrestrial life.

There are approximately 8,600 species of **birds**, which makes them second only to the bony fishes in vertebrate diversity. Yet birds are generally very similar to each other in many ways, because flight requires a very specialized anatomy.

All birds have wings, which are modifications of the front limbs, and hind limbs that may be suitable for walking, perching, swimming, grasping prey, and scratching. All birds have horny beaks and all lay eggs. In addition, all birds have a unique feature shared by no other animals: feathers. Birds evolved from a line of reptiles, and feathers evolved from reptilian scales. Birds still have scales on parts of their bodies, especially the feet. In addition to their essential role in flight, feathers provide insulation from the elements.

Birds eat a variety of animal foods, including insects and other invertebrates, such as worms, and vertebrates such as fish and frogs, reptiles, and small mammals. They also eat plant foods such as seeds, fruit, and nectar. Their beaks are adapted for their food sources. The long probing bill of the hummingbird, the hooked bill of the hawk, and the thick seed-crushing bill of the cardinal are just a few examples.

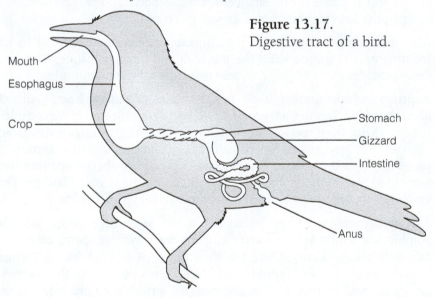

Figure 13.17.
Digestive tract of a bird.

Mouth
Esophagus
Crop
Stomach
Gizzard
Intestine
Anus

Figure 13.17 shows the digestive system of a typical bird. Compare this to the digestive system of the earthworm and the frog. Explain the similarities and differences among the animals in terms of their diets.

Like their reptilian ancestors, birds have amniotic eggs. The chicken egg is a familiar example. The yolk and the protein in the egg white supply the food for the developing embryo. The egg white also contains a great deal of water, and the whole thing is sealed into a watertight shell. The only substance that enters the egg from the environment is oxygen.

Which came first, the chicken or the egg?

The **mammals** include approximately 4,500 species ranging in size from under an inch long (the masked shrew) to 100 feet (the blue whale) and adapted for diverse lifestyles. Mammals fly, swim, climb trees, burrow underground, run on four legs, and walk upright on two legs. Note that locomotion in mammals differs from that of reptiles in that the limbs are placed more directly under the body. The characteristics shared by all mammals are bearing fur or hair and nourishing the young with milk. The young of most mammals develop in a uterus inside the mother's body, but the duck-billed platypus and the spiny anteater lay eggs. Marsupial mammals, including opossums, kangaroos, and koalas, do develop their young in a uterus, but only for a brief period. Development is completed in a pouch on the mother's body after the offspring are born.

As in other groups, many of the adaptations of mammals reflect diet. Some use specialized food resources, while others eat a wide range of foods. Herbivores, carnivores, insectivores, and omnivores are all found among the mammals. The digestive tracts of mammals also reflect their diets. In general, carnivores have relatively short digestive tracts, as you saw in the carnivorous frog. Carnivores have separate meals with time between. Herbivores, on the other hand, spend much of their day eating. What might be the reason for this difference? (Hint: Consider the composition of the diets. Do you get more calories from a salad or a hamburger?)

Like termites, vertebrates lack the enzymes necessary for digesting cellulose. And herbivores have a similar solution to the problem: gut symbionts, mainly bacteria. Herbivores have very long digestive tracts to allow time and space for the extra processing steps needed.

Since almost all mammals have teeth, comparison of teeth is a good way to study feeding adaptations. Mammals have four types of teeth (Figure 13.18). Incisors are used primarily for snipping or biting off a piece of food, the sharp canines are used to pierce, premolars shear or slice, and molars crush and grind. Depending on the animal's diet, certain teeth are emphasized and modified, while other types may be absent.

Figure 13.18.
Types of teeth in a human.

Incisors

Canine

Premolars

Molars

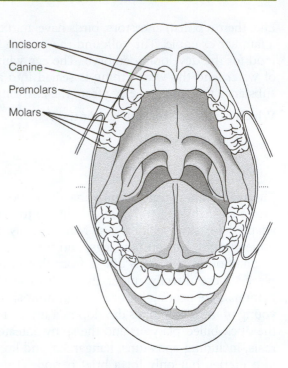

Procedure

View the skulls or photographs on display. For each specimen, describe the teeth and state your inferences about the animal's diet.

Questions for Review

1. Examine the "unknown" animals on display. Decide which phylum each animal belongs to and explain what characteristics you used to make your decision.

2. Why are sponges named pore bearers?

3. Bivalves have little sensory ability and heavy shells, while cephalopods have well-developed senses and reduced shells. How are these characteristics adaptive for each group?

4. Compare the echinoderms with the cnidarians. Consider habitat, symmetry, locomotion, defense, and means of obtaining food.

5. Name the three phyla that contain animals called worms. Discuss the differences among the digestive tracts of these three phyla.

6. Concentrating on the evolutionary transition from amphibians to reptiles, discuss the adaptations that allow vertebrates to be successful in terrestrial habitats.

7. Using as examples at least two bilaterally symmetrical organisms and two radially symmetrical organisms, discuss the adaptive value of bilateral versus radial symmetry.

8. Choose three animals from different phyla and describe their adaptations for obtaining food. Explain how these adaptations are suited to each animal's environment.

9. Choose three animals from different phyla and compare their means of digesting food. Include in your answer a comparison of the efficiency of each digestive system.

Digestion

Introduction

Where do the molecules in our bodies come from? Unlike plants, we cannot turn carbon dioxide from the air into carbohydrates, nor can we absorb other necessary nutrients from the environment. We and other animals ingest the raw materials for our molecules in the form of food. In the process of **digestion,** the food items are broken down into their component parts. They are then changed and repackaged as molecules that fit our immediate needs. They may also be stored for later use. This is analogous to the way a car is stripped by professional thieves: Its parts are redistributed and are no longer recognizable as parts belonging to the original car.

Most of the food that we eat consists of the biological macromolecules that were studied in Lab Topic 3 (Macromolecules): proteins, carbohydrates, lipids, and nucleic acids. In this lab topic, you will learn how macromolecules are broken down into their subunits so that they can be absorbed into the cells of the body. You will then design and perform an experiment using one of the methods you have learned for investigating food molecules.

 Wear safety goggles while performing these experiments.

Outline

EXERCISE 14.1
Digestion of Proteins by Trypsin

Objectives

After completing this exercise, you should be able to

1. Explain the purpose of protein digestion.
2. Identify the enzymes involved in protein digestion.
3. Explain how the procedure demonstrated in this exercise is used to determine whether protein digestion has occurred.

Introduction

The proteins we eat can be neither absorbed nor used by the body directly. They must be completely dismantled into their subunits, amino acids. We then use these amino acids to assemble our own proteins. Thus, even though we might eat quantities of muscle (meat) containing the proteins actin and myosin, these protein molecules do not necessarily become actin and myosin in the muscles in our bodies. Their amino acids are reassigned to whatever proteins are needed at the moment, which might be keratin for hair, hemoglobin for red blood cells, an enzyme in the aerobic respiration pathway, or any one of thousands of other proteins. Nine of the 20 amino acids that we need to make proteins can be obtained only by breaking down proteins from food. These are called the nine essential amino acids. We can synthesize the other 11 from simpler precursor molecules.

Chemical digestion of protein begins in the stomach. The pH of the fluids in the stomach is extremely acidic, and the primary enzyme for protein digestion in the stomach, pepsin, works best in those conditions. Further protein digestion is carried out in the small intestine by **trypsin**, the enzyme that you will use for the following procedure. Trypsin is produced in the **pancreas**, a gland that secretes hormones as well as digestive enzymes, and then exported to the small intestine. Unlike pepsin, trypsin functions most effectively under the slightly alkaline conditions of the small intestine. The enzymes that digest proteins are called **proteases.**

Method

In the following procedure you will use the enzyme trypsin, a protease present in an extract made from the pancreas, to digest gelatin.

Gelatin is a protein obtained from sources such as animal hooves and hides. It is capable of taking up large amounts of water, which cause it to swell and set in the familiar Jell-O fashion. If the gelatin is digested—for example, if trypsin has begun to break down the protein molecules into their subunits—the gelatin's structure is destroyed. Without the proper structure, the gelatin will not gel. Figure 14.1 illustrates this test.

If you were to design an experiment using this technique, what dependent variable would you be measuring?

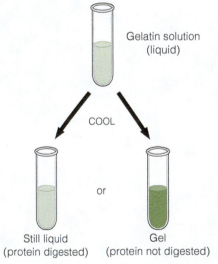

Gelatin solution (liquid)

COOL

or

Still liquid (protein digested)

Gel (protein not digested)

Figure 14.1.
Gelatin test for protease activity. The gelatin will gel only if its protein structure is intact. Digested protein will not gel.

How is this dependent variable related to the activity of trypsin?

What factors might be used as independent variables?

Predict the outcomes in Tubes 1 and 2. Justify your predictions.

Procedure

1. Get two test tubes and label them 1 and 2 with a wax marker.
2. Use a pipet to measure 2 mL of water into Tube 1.
3. Use a pipet to measure 2 mL of pancreatic extract (trypsin) into Tube 2.
4. Use a pipet or a 10-mL graduated cylinder to measure 10 mL of liquid gelatin into each tube. Swirl to mix.
5. Place the test tubes in ice for 20–30 minutes.

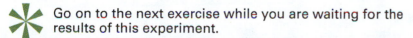   Go on to the next exercise while you are waiting for the results of this experiment.

6. After 20–30 minutes, place your thumb over the top of each tube and invert it to test whether or not the gelatin has set.
7. Record your results below and explain what happened in each tube.
    Tube 1:

    Tube 2:

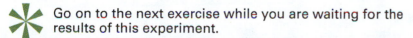   After you have obtained results for the experiment, be sure to clean your glassware thoroughly.

EXERCISE 14.2
Digestion of Lipids by Lipase

Objectives

After completing this exercise, you should be able to

1. Explain the purpose of lipid digestion.
2. Explain the difference between emulsion and digestion and explain why emulsion of lipids is necessary.
3. Identify the enzyme involved in lipid digestion.
4. Explain how the procedure in this exercise is used to determine whether lipid digestion has occurred.

Introduction

A fundamental characteristic of lipids is that they are insoluble in water. For example, salad dressing made from oil and vinegar will separate into two layers. Why does this property of lipids pose a problem for the digestive system? (Hint: What is the primary ingredient in the beverages we drink?)

Lipid digestion must begin by making the molecules more compatible with water so that the digestive enzymes have access to them. This is accomplished by breaking up the lipid into small droplets, which are then distributed throughout the water. This type of mixture is called an **emulsion.** Immediately before pouring out salad dressing, you shake the bottle. The separation problem is solved only temporarily, however. In a few minutes, the oil droplets have again joined up with each other and the oil and aqueous vinegar have separated. A permanent emulsion can be established by using an emulsifying agent to keep the oil droplets dispersed. In salad dressing, mustard may be added to stabilize the oil-and-vinegar emulsion. Egg yolks are also emulsifiers. For example, mayonnaise is an emulsion containing oil, lemon juice, and egg yolks. In the digestive process, **bile,** which is produced by the liver and stored in the gall bladder, emulsifies lipids.

Once lipids are emulsified they are broken down into their subunits by enzymes called **lipases.** The subunits can then be absorbed into the body. Lipases are another type of digestive enzyme that is produced in the pancreas and secreted into the small intestine. Lipid emulsification and digestion then take place in the small intestine, and the subunits are absorbed into the blood.

Fats, which constitute a major subclass of lipids, are composed of a glycerol molecule linked to fatty acids. Most fats found in nature have three fatty acids, so they are called triglycerides. There are also fats that contain one or two fatty acids linked to a glycerol molecule. Enzymatic digestion of fats releases the fatty acids. As weak acids, the fatty acids affect the pH of a solution.

Method

In the following procedure you will use the acidic property of fatty acids to determine whether digestion of fats has occurred. You will mix vegetable oil, a fat, with pancreatic extract, which contains lipases as well as trypsin. If fats are being digested, will the pH of the solution increase or decrease?

You will use phenolphthalein as a pH indicator. NaOH (sodium hydroxide) is added to make each solution strongly basic. Phenolphthalein turns pink at pH 10. As digestion occurs and fatty acids are released, the phenolphthalein will become colorless, indicating that the solution is now only slightly basic (approximately pH 8).

Examine Table 14.1, which shows how the experiment is set up, and answer the questions that follow.

Table 14.1
Amounts of Reactants Used in Lipid Digestion Procedure

	Tube		
	1	2	3
Vegetable oil (mL)	1	1	1
Bile (mL)	0	1	1
Water (mL)	2	1	6
Phenolphthalein (dropperful)	1	1	1
Pancreatic extract (mL)	5	5	0

What hypothesis is being tested in this experiment?

What dependent variable is measured by the procedure that is used?

How is this dependent variable related to the activity of lipase?

What is the independent variable in this experiment?

What other factors might affect lipase activity?

What control treatments are used in this experiment?

Predict the outcome of this experiment and justify your prediction.

Procedure

1. Get three test tubes and label them 1, 2, and 3 with a wax marker.
2. Use a pipet to measure 1 mL of vegetable oil into Tubes 1, 2, and 3.
3. Use a pipet to measure 1 mL of bile into Tubes 2 and 3.
4. Use a pipet to measure water into each tube, as shown in Table 14.1.
5. Add 1 dropperful of phenolphthalein from a dropping bottle to each tube.
6. Using a pipet, add 5 mL pancreatic extract to Tubes 1 and 2.
7. Cover each tube tightly with Parafilm, put your thumb over the top, and shake vigorously ten times.

 NaOH is a caustic solution. Avoid contact with your skin or clothing. Inform your instructor immediately if a spill occurs.

8. Remove the Parafilm and put 2–3 drops of NaOH from a dropping bottle into each tube. Replace the Parafilm and shake to mix the solution. Add drops of NaOH until the solution stays pink.
9. Record the color in each test tube in Table 14.2 at time 0.

 Release of fatty acids will change the solution from pink to colorless. However, the bile solution in Tubes 2 and 3 will give them a pale yellow cast throughout the experiment.

Table 14.2
Results of Lipid Digestion Procedure

	Tube		
	1	**2**	**3**
Time (min)	**Oil + Lipase**	**Oil + Bile + Lipase**	**Oil + Bile**
0			
1			
2			
3			
4			
5			
20			
30			

10. For the next 5 minutes, record the color in each tube at 1-minute intervals in Table 14.2.

11. Record the colors again after 20 minutes.

 While you are waiting for the results from this procedure, go back to Exercise 14.1 and observe the results. Then begin Exercise 14.3.

12. Record the colors again after 30 minutes, if time permits.

When you have obtained the results of this experiment, be sure to clean your glassware thoroughly. Soaps and detergents are also emulsifying agents, breaking grease up into small particles that are more easily carried away from your dishes. Please keep this in mind when you clean up your glassware.

Explain the color change or lack of color change in each tube.

Tube 1

Tube 2

Tube 3

EXERCISE 14.3
Digestion of Starch by Amylase

Objectives

After completing this exercise, you should be able to

1. Explain the purpose of carbohydrate digestion.
2. Explain how the procedures in this exercise are used to determine whether carbohydrate digestion has occurred.

Introduction

Sugars are simple carbohydrates. Polysaccharides such as starch and cellulose are complex carbohydrates. What subunits compose these two polysaccharides?

The ultimate function of carbohydrate digestion is to convert all carbohydrate molecules into monosaccharide subunits that can be absorbed into the blood and ultimately into body cells. The giant polysaccharides obviously must be chopped up into their component subunits, but even disaccharides such as sucrose (table sugar) and lactose (milk sugar) must be converted to monosaccharides for absorption.

Cellulose, which composes the cell walls of plants, is a large polysaccharide and a component of many plant foods we eat, but our digestive systems cannot break cellulose molecules down into their component sugars. Because it is indigestible, cellulose is the largest component of dietary fiber. It adds bulk to the intestinal contents and thus helps the remains of a meal pass through the digestive system more quickly.

On the other hand, starch is a major source of nutrients. Digestion of starch begins in the mouth, where salivary **amylase** initiates the process of clipping the huge polysaccharide into smaller strands, then into disaccharide maltose molecules, and finally into its monosaccharide glucose subunits. Like trypsin and lipase, amylase is also secreted by the pancreas into the small intestine. In the small intestine, starch digestion that was begun in the mouth is completed, and the monosaccharides are absorbed into the blood.

Method

For this procedure you will use pancreatic extract as a source of amylase. Since the activity of digestive enzymes is affected by pH, you will add HCl (hydrochloric acid) to one of the experimental tubes to determine its effect on amylase activity.

Starch is added to the experimental tubes as a substrate for amylase. You will use iodine reagent, I_2KI (iodine-potassium iodide), which turns dark blue in the presence of starch, to determine whether starch digestion has occurred. If no starch is present, the solution remains yellow. If the starch has been partially digested, a red-brown color will result.

Examine Table 14.3 to see how the experiment has been set up, then answer the questions that follow.

Table 14.3
Amounts of Reactants Used in Starch Digestion Procedure

	Tube			
	1	**2**	**3**	**4**
Pancreatic extract (mL)	0	1	1	1
Starch (mL)	5	5	5	0
0.5% HCl (dropperful)	0	0	1	0
Water (mL)	2	1	0	6

What hypothesis is being tested in this experiment?

What dependent variable is measured by this procedure?

How is this dependent variable related to the activity of amylase?

What is the independent variable in this experiment?

What other factors can affect amylase activity?

What control treatments are used in this experiment?

Predict the outcome of this experiment and justify your prediction.

Procedure

1. Get four test tubes and label them 1, 2, 3, and 4 with a wax marker.
2. Use a pipet to measure 5 mL starch solution into Tubes 1, 2, and 3.
3. Add 1 dropperful of 0.5% HCl from a dropping bottle to Tube 3.

> ⚠ HCl is a caustic solution. Avoid contact with your skin or clothing. Inform your instructor immediately if a spill occurs.

4. Using a pipet, add 2 mL water to Tube 1, 1 mL water to Tube 2, and 6 mL water to Tube 4 (see Table 14.3).
5. Using a pipet, add 1 mL of pancreatic extract to Tubes 2, 3, and 4.
6. Cover each tube tightly with Parafilm and invert once to mix.
7. Use a wax marker to label a second set of test tubes 1, 2, 3, and 4. Draw one pasteur pipetful of solution out of each tube in the first set and put it into the corresponding tube in the second set.
8. Put a few drops of iodine reagent from a dropping bottle into each tube in the second set.
9. Record the results as + (starch present) or − (no starch) in Table 14.4.

 A blue color indicates that undigested starch is present. Yellow means that no starch is present.

Table 14.4
Results of Starch Digestion Experiment

Iodine Test for Starch	Tube			
	1	2	3	4
	Starch	Amylase + Starch	Amylase + Starch + HCl	Amylase
Initial				
Final				

10. Wait at least 5 minutes for digestion to occur in the first set of tubes. While you are waiting, label a third set of test tubes 1, 2, 3, and 4.

11. After 5 minutes, test again for starch in each tube by drawing one pasteur pipetful of solution out of each tube in the first (experimental) set and putting it into the corresponding tube in the third set. Then add a few drops of iodine reagent to each tube in the third set.

12. Record the results in Table 14.4.

 After you have obtained results for the experiment, be sure to clean your glassware thoroughly.

What is a positive reaction to this test?

If a solution that contains amylase (pancreatic extract) tests positive for starch after 5 minutes, what does this indicate about amylase activity?

Is your hypothesis supported or proven false by the results? Explain.

EXERCISE 14.4
Designing an Experiment

Objective

After completing this exercise, you should be able to

1. Design an original experiment to investigate some factor that affects the process of digestion.

In Exercises 14.1, 14.2, and 14.3 you learned techniques for investigating three digestive enzymes that are contained in an extract made from the pancreas: trypsin, lipase, and amylase. In Exercises 14.4 and 14.5 your lab team will design an experiment using one of these methods, perform your experiment, and present and interpret your results. You may review the factors that you listed as possibly affecting the activity of trypsin, lipase, and amylase to help you decide on an independent variable for your investigtion.

The following materials will be provided for your team.

For Trypsin Experiment

test tubes	ice chest with ice
test tube rack	pancreatic extract
10-mL graduated cylinder	2% gelatin solution
5-mL pipets with pi-pump	

For Lipase Experiment

test tubes	0.1*N* NaOH
small Parafilm squares	vegetable oil
pancreatic extract	5-mL pipets with pi-pump
phenolphthalein solution	

For Amylase Experiment

test tubes	pancreatic extract
1-mL pipets with pi-pump	starch solution
5-mL pipets with pi-pump	iodine reagent

Your instructor will tell you what additional materials will be available.

Describe your experiment below.

Question or Hypothesis

Dependent Variable

Independent Variable

Explain why you think this independent variable will affect the dependent variable.

Control Treatment(s)

Replication

Brief Explanation of Experiment

Predictions

What results would support your hypothesis? What results would prove your hypothesis false?

Method

You may find it helpful to construct a table showing the contents of each reaction tube.

Design a Table to Collect Your Data

List Any Additional Materials You Will Require

E X E R C I S E   1 4 . 5
Performing the Experiment and Interpreting the Results

Objectives

After completing this exercise, you should be able to

1. Perform the experiment your lab team designed.
2. Present and interpret the results of your experiment.

Before you do the experiment, be sure that everyone on your lab team understands the techniques that will be used. You may divide up the tasks before you begin work.

Be thorough in collecting data. Don't just write down numbers; record what they mean as well. Don't rely on your memory for information that you will need when reporting on your experiment later! If you have any questions, doubts, or problems during the experiment, be sure to write them down, too.

Results

Before you begin to prepare your results for presentation, decide on the best format to use. Remember, you want to give the reader a clear, concise picture of what your experiment showed. Refer to the data presentation section of Appendix A (Tools for Scientific Inquiry) for help. If you are drawing graphs, use graph paper. Complete your tables and/or graphs before attempting to interpret your results.

Write a few sentences below *describing* the results (don't explain why you got these results or draw conclusions yet).

Discussion

Look back at the hypothesis or question you posed in this experiment. Look at the graphs or tables of your data. Do your results support your hypothesis or prove it false? Explain your answer, using your data for support.

Did your results correspond to the prediction you made? If not, explain how your results are different from your expectations and why this might have occurred.

Describe how your data are supported by information from other sources (for example, textbooks or other lab teams working on a similar problem).

If you had any problems with the procedure or questionable results, explain how they might have influenced your conclusion.

If you had an opportunity to repeat and extend this experiment to make your results more convincing, what would you do?

Summarize the conclusion you have drawn from your results.

Questions for Review

1. As you are happily munching on your Big Mac, your mind wanders back to bio lab and what you learned about digestion. You are moved to describe for your friends the fate of the two all-beef patties, special sauce, lettuce, cheese, pickles, and onions on a sesame seed bun. How is each of these ingredients digested and in what form do they finally reach your bloodstream? (Also include any indigestible parts of the food.)

   Two all-beef patties

   Special sauce

   Onions, lettuce, pickles

   Cheese

   Sesame seed bun

2. Fungi lie in contact with their food sources (for example, a slice of bread) as they grow. The fungus secretes enzymes that break down large food molecules, and the fungal cells then absorb the products of

that digestion. Compare this process to the digestive processes described in this lab topic.

3.   Describe a test to determine whether gelatin has been digested.

4.   An unknown combination of macromolecules is mixed with pancreatic extract. Samples are tested at the beginning of the experiment and after 30 minutes. The following results are observed.

Phenolphthalein test: Initially red, colorless after 30 minutes.

Iodine test: An initial sample is blue. A sample taken after 30 minutes is yellow.

What macromolecules were present in this sample before digestion? How do you know?

What molecules are present in the sample after 30 minutes? How do you know?

Could there have been macromolecules in the sample whose presence you didn't detect? Explain.

Acknowledgments

Procedures for the enzyme assays were adapted from the following sources:

Armstrong, W. D., and C. W. Carr. *Physiological Chemistry Laboratory Directions,* 3rd ed. Minneapolis: Burgess Publishing, 1963.

Dotti, L. B., and J. M. Orten. *Laboratory Instructions in Biochemistry,* 8th ed. St. Louis: C. V. Mosby, 1971.

Oser, B. L., ed. *Hawk's Physiological Chemistry,* 14th ed. New York: McGraw-Hill, 1965.

Circulation

Introduction

In order to carry on all the functions necessary for life, animal cells must receive nutrients and oxygen from their environment. They must also be able to rid themselves of metabolic wastes, including carbon dioxide. In the least complex animals, such as sponges, cnidarians, flatworms, and roundworms, most cells either are in direct contact with the external environment or are only one cell away. Therefore, these animals can obtain oxygen and nutrients directly into their cells by diffusion. In more complex animals, however, a circulatory system is needed to transport nutrients and oxygen to all cells in the body and to remove metabolic wastes. Blood is the tissue responsible for these transport functions. In this lab topic you will learn techniques for investigating two aspects of the human circulatory system, which transports oxygen, nutrients, and other substances throughout the body. You will also design and perform your own experiment using these techniques.

Outline

EXERCISE 15.1
Anatomy of the Mammalian Circulatory System

Objectives

After completing this exercise, you should be able to

1. Explain the roles of the heart, arteries, veins, and capillaries in the circulatory system.
2. Identify the structures of the mammalian heart and indicate the direction of blood flow into, through, and out of the heart.

3. Distinguish between arteries and veins.
4. Explain how the structure of arteries is related to their function.
5. Explain how the structure of veins is related to their function.
6. Explain how the structure of capillaries is related to their function.

Introduction

Blood is distributed from the heart to the lungs and body tissues by a system of blood vessels (Figure 15.1). **Arteries** carry blood away from the heart. Major arteries branch into smaller arteries called **arterioles** that narrow into capillaries that are in close contact with the tissues. **Capillaries** are thin-walled vessels that permit exchange of gases and other substances between the blood and the tissues. Capillaries merge into **venules**, small vessels that connect the capillaries with veins. The **veins** carry blood back to the heart to complete the circuit.

Figure 15.1.
Components of the mammalian circulatory system.

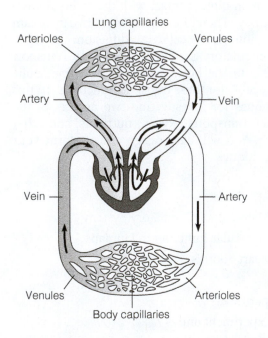

In this exercise you will identify the structures of the mammalian heart using a pig heart as an example. Look at the pig heart provided for this exercise and identify the structures that are listed in the following paragraphs in boldface type. Label these structures on the diagram of the heart (Figure 15.2). Draw arrows to indicate the direction of blood flow into, through, and out of the heart.

Note that in Figure 15.2 you are viewing the heart as if you were facing the animal: The right side of the diagram is the left side of the heart. Hold the pig heart so that it is in the same position as the diagram.

⚠️ **Wear latex gloves when you are handling the heart.**

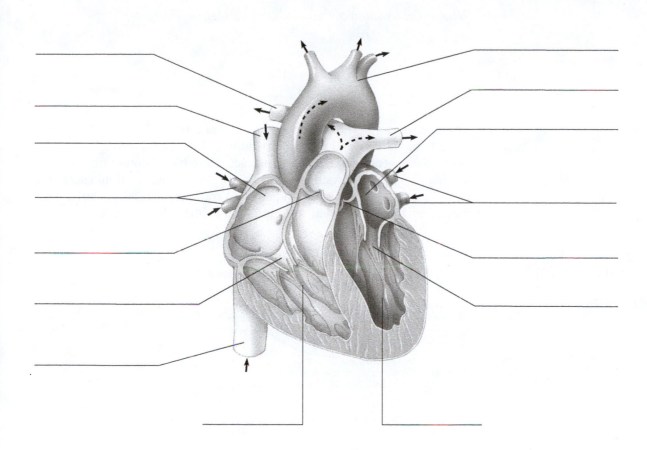

Figure 15.2.
Mammalian heart. The human heart, shown above, is very similar to the
pig heart that you will observe in lab.

Blood returning to the heart from the body enters the **right atrium** via the
two vessels called the **venae cavae** (singular: vena cava). The superior
vena cava carries blood from the upper portions of the body and the inferior vena cava brings blood from the lower body. From the right atrium,
blood passes through the **right atrioventricular (AV) valve** to the **right
ventricle**, which pumps it through the **pulmonary artery** to the lungs.
There is a **semilunar valve** between the right ventricle and the pulmonary
artery. Blood returns to the **left atrium** from the lungs via the **pulmonary
veins.** The blood then passes through the **left atrioventricular valve** to
the **left ventricle**, which pumps it out to the body through the **aorta**.
Notice that the muscle of the left ventricular wall is thicker than the right
ventricular wall. Blood leaving the left ventricle must be pumped farther
than blood leaving the right ventricle. There is another semilunar valve
between the left ventricle and the aorta.

What is the function of the atrioventricular valves?

What is the function of the semilunar valves?

The arteries can be viewed as muscular, elastic tubes that offer little resistance to blood flow and provide pressure to drive blood through the body. When the ventricles of the heart contract, more blood flows into the arteries than leaves them (systole, Figure 15.3). The volume of fluid causes the elastic arterial walls to bulge. When the ventricles relax, the arterial walls recoil and give the blood an additional push (diastole, Figure 15.3).

Figure 15.3.
Movement of blood through arteries.

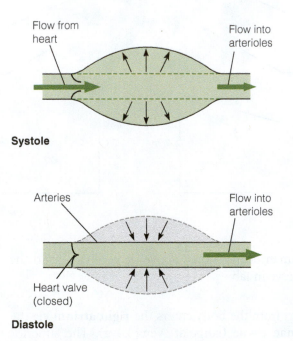

By the time blood reaches the venules to begin its journey back to the heart, blood pressure is very low. The veins have large diameters and provide little resistance to blood flow. They also have thinner walls, which are able to stretch more than arterial walls; therefore, veins can hold larger volumes of blood with less increase in pressure than arteries. Venous pressure can be increased by contraction of the smooth muscle in the walls of the veins. It is also increased when veins are squeezed by the surrounding skeletal muscle.

Procedure

1. View the microscope slide of an artery and a vein in cross section (the artery has a thicker wall and is more circular in shape than the vein).

2. Sketch the artery in the margin of your lab manual. Notice the thick middle layer of elastic connective tissue and smooth muscle in the artery. This layer of tissue provides resistance to stretching, thus keeping the blood under pressure.

3. Sketch the vein in the margin.

What is the principal difference between the two vessels?

The function of **capillaries** is to provide an interface between the blood and the other body tissues. The diameter of capillaries is often not much larger than the diameter of a red blood cell, and the cells must travel in single file. Capillary walls are only one cell layer thick. Explain how this structure serves capillary function.

EXERCISE 15.2
Measuring Cardiovascular Function in Humans

Objectives

After completing this exercise, you should be able to

1. Determine pulse rate for an individual.
2. Determine blood pressure for an individual.
3. Describe the cardiac cycle and how it relates to heart sounds and blood pressure.
4. Explain how and why hypertension, atherosclerosis, and arteriosclerosis affect circulation.

Activity A: Pulse Rate and Heart Sounds

The most accessible measure of cardiac activity is **pulse rate**. As the blood is pumped, the intermittent pressure can be felt in the arteries, especially at the wrist and throat. (See Figure 15.4.) Counting the pulse rate tells us the number of heart beats per minute.

Procedure for Pulse Rate

1. Find your pulse as shown in Figure 15.4. Use your fingers rather than your thumb, since there is a slight pulse in the thumb.
2. Count your pulse for 15 seconds _____ .
3. Multiply this figure by four to determine the number of beats per minute: _____ . This is your resting pulse rate.

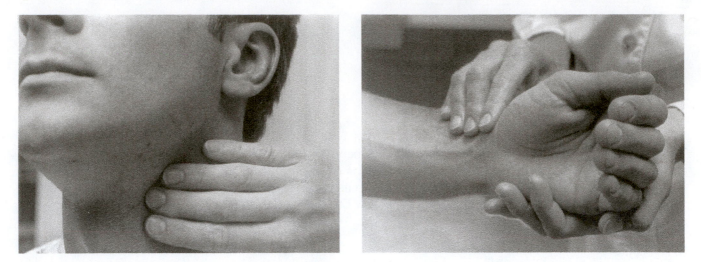

Figure 15.4.
Measuring pulse rate.

Contraction of the ventricles propels blood through the arteries, creating the pulse. The period from the beginning of one ventricular contraction to the beginning of the next is called the **cardiac cycle.** The ventricular contraction itself is referred to as **systole,** and it is followed by a period of relaxation called **diastole.** You will use a stethoscope to listen to the events of the cardiac cycle.

Procedure for Heart Sounds

1. Clean the ear pieces of a stethoscope with alcohol.
2. Place the chestpiece of the stethoscope to the left of your (or your partner's) breastbone and listen to the heart.
3. You should hear a low-pitched "lub" sound followed by a higher-pitched, more forceful "dup." "Lub" is the sound of the atrioventricular (AV) valves closing. Would this mark the beginning of systole or diastole? (Refer to Figure 15.2 if you have forgotten the location of the AV valves.)

4. Now listen for the "dup" sound. "Dup" is heard when the semilunar valves close. Is "dup" the beginning of systole or diastole?

Normally you hear only the vibrations made by the valve closures; the blood flowing through the heart is not heard. A **heart murmur** means that the rush of blood is audible. This may occur because the valves fail to close completely, allowing backflow; there may be a small hole in one of the walls that divide the heart chambers; or the valve opening may be abnormally narrow. In any case the murmur is a result of faster, more turbulent blood flow than is normal.

Activity B: Blood Pressure

The flow of blood into, through, and out of the heart is driven by pressure differences created by the cardiac contractions. Once the blood is out in the body, it is pushed along by **arterial pressure** (review Figure 15.3). What we call **blood pressure** is actually arterial pressure. Maximum arterial pressure coincides with ventricular contraction (systole). Minimum pressure (diastole) occurs before the next ventricular contraction. Note that although the pressure is lower during ventricular relaxation, it is well above zero. Blood flow does not stop between ventricular contractions.

You can present blood pressure data either by graphing systolic pressure and diastolic pressure on the same axes (Figure 15.5a) or by plotting the difference between the systolic and diastolic pressures (Figure 15.5b), which is called **pulse pressure.** It is useful to show pulse pressures when comparing one group of subjects to another or when comparing data taken from a subject at different times.

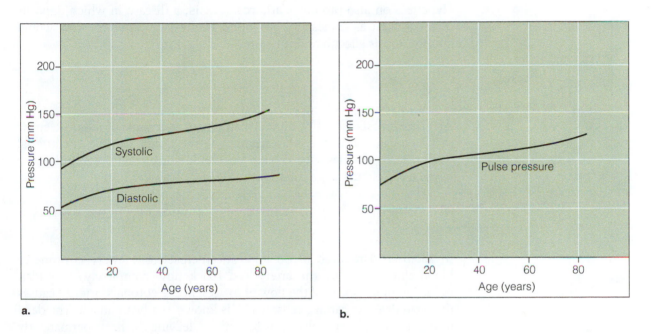

a.   b.

Figure 15.5.
Graphical presentation of blood pressure data. (a) Changes in systolic and diastolic pressures with age. (b) Changes in pulse pressure with age.

Systolic pressure normally ranges from 108 to 140 mm Hg (millimeters of mercury, a measure of pressure) and normal diastolic pressure ranges from 60 to 90 mm Hg for 20–24-year-olds. Average values for males are slightly higher than for females, and blood pressures increase for both sexes with age (see Figure 15.5).

High blood pressure, or hypertension, is an increased resistance to blood flow in the arteries; that is, more force is required to pump blood through the arteries. Generally **hypertension** is defined as blood pressure greater than 140/90 mm Hg. High blood pressure can be dangerous because an artery might rupture under the increased pressure. The two most important types of damage to arteries from hypertension are to vessels in the brain and kidneys. High blood pressure also forces the heart muscle to work harder to pump blood to the body. Like all muscle, the heart requires oxygen to generate the ATP necessary for contraction. Oxygen and nutrients are brought to the heart by the coronary arteries. The increased workload of the muscle must be supported by an increase in oxygen. Angina pectoris (chest pain) results from an oxygen deficiency.

What circumstances might trigger angina in a person who has hypertension?

Why does exercise increase the heart rate?

Hypertension also promotes **atherosclerosis,** a disease in which deposits of lipids such as cholesterol are found in the large arteries. How would atherosclerosis affect blood pressure?

Arteriosclerosis (literally, arterial hardening) is a form of atherosclerosis in which calcium depositions accompany the lipids, causing the arterial walls to lose their elasticity. What effect would you expect this condition to have on the circulatory system? (Hint: Review Figure 15.3.)

In addition to reduced blood flow, an atherosclerotic artery is prone to blood clots that plug the artery and block flow completely. A sudden reduction or stoppage of the flow of oxygen in the coronary arteries causes the rapid death of cardiac cells, which is known as a heart attack. The dead muscle cells are later replaced by scar tissue, leaving the heart permanently weakened.

In the following procedure you will take your partner's blood pressure using a stethoscope, a blood pressure cuff, and a device that measures the pressure in terms of millimeters of mercury (mm Hg). You will take one blood pressure reading with the subject lying down and one reading with the subject standing. What hypothesis is being tested with these data?

What is the independent variable?

What is the dependent variable?

Why do you think this independent variable might affect the dependent variable?

Predict the outcome of the experiment in terms of your hypothesis.

 Read the entire procedure before you start.

Procedure

Figure 15.6 shows where to place the cuff and stethoscope. It is helpful to locate the brachial artery first by feeling for it with your fingers. Figure 15.7 relates the procedure of taking blood pressure to the events of the cardiac cycle and illustrates the variations in arterial pressure. The circled numbers will help you understand how these variations are related to the cardiac cycle.

 Work with a lab partner. One of you should be the subject and the other should take the blood pressure readings.

1. Take the first blood pressure reading with the subject lying down. The subject should lie still for 2–3 minutes before the blood pressure is taken.
2. Wrap the cuff around the arm just above the elbow. It should be just loose enough so that there is room for two fingers between the cuff and the arm.
3. Close the valve on the bulb by turning it all the way clockwise.
4. Locate the brachial artery (see Figure 15.6) and hold the stethoscope in place over it.

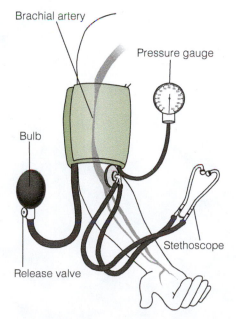

Figure 15.6.
Position of blood pressure cuff and stethoscope.

Figure 15.7.
Relation of arterial and cuff pressures.

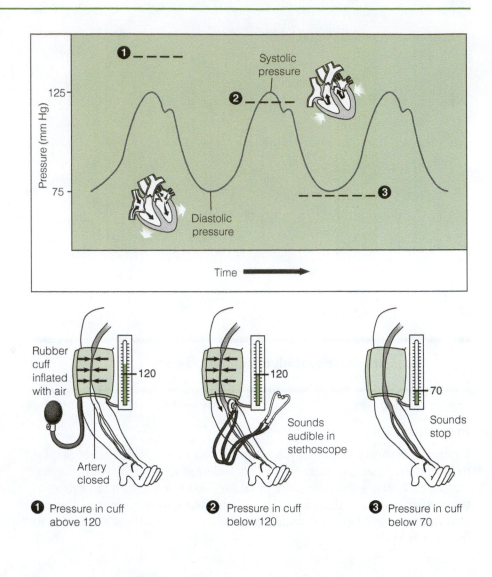

5. Inflate the cuff to a pressure higher than systolic pressure (① on Figure 15.7); 160 mm Hg should be adequate. The cuff presses on the arteries, shutting off blood flow to the lower arm.

 Do not keep the cuff inflated for more than one minute. If you can't complete the procedure, deflate the cuff and wait a few minutes before you try again.

6. *Very slowly* release the pressure by turning the valve slightly counter-clockwise. Listen through the stethoscope. When the cuff pressure is slightly below systolic pressure, the artery begins to open and blood starts to flow through ②. As in a heart murmur, this blood flow is more audible because constriction of the artery causes turbulent, high-velocity flow. When you hear the first intermittent tapping noises, record the pressure: _____ mm Hg. This is the systolic pressure.

7. As the cuff deflates, the sounds are continuous but have become muffled ③. When the cuff pressure is slightly below diastolic

pressure, blood flow through the artery is normal and the sounds disappear. Record the point at which you stop hearing the sounds: _____ mm Hg. This is the diastolic pressure.

To summarize: Systolic pressure is recorded when sounds are first heard. Diastolic pressure is recorded when sounds disappear.

⚠️ **If you are not successful in determining your partner's blood pressure after three attempts, don't try again. Repeatedly collapsing the arteries can be harmful.**

---

8. Take the subject's blood pressure again while he or she is standing up. Record systolic/diastolic pressure: _____ / _____ mm Hg.

The instructor will compile class data. Decide on an appropriate method of presentation and interpret your results.

Was your hypothesis proven false or supported by the results? Use data to support your answer.

What other independent variables might affect blood pressure and/or pulse rate?

E X E R C I S E   1 5 . 3
Designing an Experiment

Objective

After completing this exercise, you should be able to

1. Design an original experiment to investigate some factor that affects pulse rate and/or blood pressure.

In Exercises 15.1 and 15.2 you learned how to measure pulse rate and blood pressure. In Exercises 15.3 and 15.4 your lab team will design an experiment using one of these methods, perform your experiment, and present and interpret your results. You may review the independent variables you listed at the end of Exercise 15.2 before beginning to design your experiment.

Describe your experiment below. To increase the number of data points you can get, you may need to ask other lab teams to participate in your experiment.

The following materials will be supplied for your group:

apparatus to measure blood pressure

stethoscope

Hypothesis

Dependent Variable

Independent Variable

Explain why you think this independent variable will affect pulse rate and/or blood pressure.

Control Treatment(s)

Replication

Brief Explanation of Experiment

Predictions

What results would support your hypothesis? What results would prove your hypothesis to be false?

Method

Design a Table to Collect Your Data

List Any Additional Materials You Will Require

E X E R C I S E   1 5 . 4

Performing the Experiment and Interpreting the Results

Objectives

After completing this exercise, you should be able to

1. Perform the experiment your lab team designed.
2. Present and interpret the results of your experiment.

Before you do the experiment, be sure that everyone on your lab team understands the techniques that will be used. You may want to divide up the tasks before you begin work.

Be thorough in collecting data. Don't just write down numbers; record what they mean as well. Don't rely on your memory for information that you will need when reporting on your experiment later! If you have any questions, doubts, or problems during the experiment, be sure to write them down, too.

Results

Before you begin to prepare your results for presentation, decide on the best format to use. Remember, you want to give the reader a clear, concise picture of what your experiment showed. Refer to the data presentation section of Appendix A (Tools for Scientific Inquiry) for help. If you are drawing graphs, use graph paper. Complete your tables and/or graphs before attempting to interpret your results.

Write a few sentences *describing* the results (don't explain why you got these results or draw conclusions yet).

Discussion

Look back at the hypothesis you posed in this experiment. Look at the graphs or tables of your data. Do your results support your hypothesis or prove it false? Explain your answer, using your data for support.

Did your results correspond to the prediction you made? If not, explain how your results are different from your expectations and why this might have occurred.

Describe how your data are supported by information from other sources (for example, textbooks or other lab teams working on a similar problem).

If you had any problems with the procedure or had questionable results, explain how they might have influenced your conclusion.

If you had an opportunity to repeat and extend this experiment to make your results more convincing, what would you do?

Summarize the conclusion you have drawn from your results.

Questions for Review

1. When blood returns to the heart from the body, it travels into the heart via the two_____ to the _____ of the heart. From there, blood enters the_____ of the heart. Backflow into the atrium is prevented by the _____ valve. When blood returns to the heart from the lungs, it travels via the_____veins into the _____ of the heart. From there, blood enters the_____ of the heart. Backflow into the atrium is prevented by the_____valve. When the right ventricle contracts, blood leaves the heart via the

_____ artery. The_____ valve prevents backflow into the ventricle. When the left ventricle contracts, blood leaves the heart via the_____ (blood vessel). The_____ prevents backflow into the ventricle.

2. The_____ valves close when the ventricles contract. The sound heard when those valves close is_____. The ventricles contract during the _____ phase of the cardiac cycle. The _____ valves close after the ventricles contract. The sound heard when those valves close is_____. The ventricles relax during the_____phase of the cardiac cycle.

3. Make a schematic diagram of the mammalian heart, showing how the blood flows through the right and left atria, the right and left ventricles, the venae cavae, the aorta, the pulmonary veins, and the pulmonary artery. Draw arrows showing the direction of blood flow.

4. If you were shown two microscope slides, how would you be able to tell which one is an artery and which one is a vein?

5. What are the functions of arteries and how does their structure help them perform these functions?

6. What are atherosclerosis and arteriosclerosis? Why are they harmful to the circulatory system?

7. What is the function of veins? How does the structure of veins help them perform their function?

8. What is the function of capillaries? How does the structure of capillaries help them perform their function?

9. What are the two phases of the cardiac cycle?

The Sensory System

Introduction

Animal survival depends on the ability to sense and respond to stimuli in the environment. As you studied animals in Lab Topic 13 (Animal Diversity), you learned that even the simplest animals have some rudimentary means of perceiving their surroundings. For example, the planarian, a flatworm, has light-sensitive eyespots in the head region. The hydra, a Cnidarian, can chemically sense the presence of food in the water. Coupled with the ability to sense a stimulus is the need to be capable of response. The planarian avoids light in order to minimize its exposure to predators. When it senses light it uses the muscles of its body to turn away. The hydra, sensing food, waves its tentacles to capture it.

Vertebrates have much more complex nervous systems. The **sensory system,** which is part of the **peripheral nervous system,** perceives environmental stimuli. It sends information about the stimuli to the central nervous system, which consists of the brain and spinal cord. A message is then sent from the **central nervous system** to the other component of the peripheral nervous system, the **motor system,** directing a response.

The sensory system, which will be studied in this lab topic, is composed of specialized neurons, or nerve cells. These **sensory receptors** are capable of detecting specific types of stimuli such as light, touch, sound, or chemicals. They convert physical stimuli, which are one form of energy, into **action potentials,** which represent electrochemical energy. That is, the stimulus causes a change in the receptor cell's membrane that is communicated from the receptor through a chain of nerve cells to the central nervous system. The sensory receptor thus acts as a transducer, changing energy from one form such as light, touch, or sound into the electrochemical energy of an action potential. For example, sound energy that is perceived by auditory receptors in the ear is transduced into the electrochemical energy of an action potential that travels to the brain. Analogously, a loudspeaker transduces electrical energy from a radio or TV into sound energy.

Sensory receptors are specialized to receive a particular type of stimulus. In these laboratory exercises you will study **chemoreceptors,** which detect chemicals; **photoreceptors,** which sense light; and **tactile receptors,** which are sensitive to pressure and temperature.

Outline

Exercise 16.1: Chemical Senses
    Activity A: Taste Discrimination
    Activity B: Individual Differences in Taste
    Activity C: Smell Discrimination
    Activity D: Influence of Smell on Taste

EXERCISE 16.1
Chemical Senses

Objectives

After completing this exercise, you should be able to

1. Name the taste organ and the specific taste receptors and explain how taste is perceived by humans.

2. Name the five basic tastes and give examples of substances that elicit them.

3. Explain the adaptive significance of the five basic tastes.

4. Explain how it is possible that different people perceive the same substances as different tastes (or having no taste).

5. Explain how smells are perceived by humans.

6. Compare the complexity of the sense of smell with taste.

7. Discuss the influence of smell on taste.

Introduction

The ability to sense chemicals in the environment arose early in the evolution of animals. Although in most animals the chemosensors are concentrated in specific areas of the head, there are variations as well. House flies have chemosensors on their feet, enabling them to "taste" a potential food source as soon as they land on it. Besides taste receptors on the whisker-like barbels or "feelers" around the mouth, catfish also have taste buds all over their bodies. Humans receive chemical stimuli in two different ways. Molecules that are carried by water enter our mouths and are sensed by taste buds. Airborne molecules are sensed by cells in the nasal cavity.

Why is it beneficial for us to have a separate system for detection of taste and smell? What is the "specialty" of each?

The human tongue contains specialized receptor organs, the taste buds, to detect dissolved chemicals. The bumps on the tongue, called papillae, are separated by trenches, and the taste buds are located along the walls of the trenches. Particles of food that become lodged between the papillae can rot and contribute to bad breath, so oral hygienists recommend that people gently brush their tongues along with their teeth. Adults have about 10,000 taste buds; children have more. As you have probably observed, the tongue and the mucous membrane of the mouth also have receptors for pressure, temperature, and pain. These receptors are responsible for the "mouth feel" that contributes the sense of pleasure or distaste aroused by particular foods.

Figure 16.1 shows a taste bud. There are 20 to 30 taste receptor cells in each taste bud. The receptor cells are continually being replaced; their life span is about 10 days. One end of the receptor cell projects into a pore of the bud so it can contact chemical substances directly. The other end is in contact with a neuron. When the receptor is stimulated by a chemical, it transduces chemical energy into an action potential, which is transmitted to the adjoining neuron and eventually to the brain.

The olfactory receptor cells, which are responsible for the sense of smell, are located in the mucous membrane in the upper part of the nasal cavity. The receptors have cilia on one end that extend into the mucous layer (Figure 16.2). For a substance to be experienced as an odor, it must release molecules that diffuse through the air to the olfactory receptors. The molecules then dissolve in the mucous layer and interact with the receptors. In the human nose, there are as many as 40 million olfactory receptors in a space the size of a dime. That sounds like a lot, but our sense of smell is rather poor compared to other animals. Dogs, for example, have a much larger area devoted to olfactory sensors and have ten times as many receptor cells as we do. As a result, they can detect scents at much weaker concentrations.

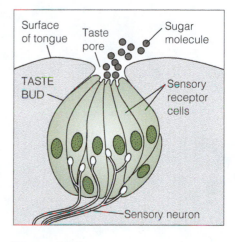

Figure 16.1.
Human taste bud. When sensory cells in the taste bud are stimulated by sugar molecules, sensory neurons transmit action potentials to the brain.

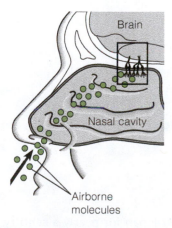

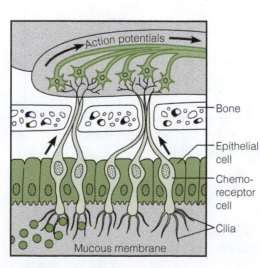

Figure 16.2.
Olfactory epithelium. The smell receptor cells sense molecules that dissolve in the mucous layer. Information from the receptor cells is sent to the brain via nerve fibers. The olfactory epithelium also includes nonreceptor epithelial cells that surround the receptors.

Activity A: Taste Discrimination

In this activity you will determine the tastes that you can identify. Each student should perform the entire procedure, but you should share materials with others working nearby.

Procedure

1. Get a small amount of the five testing solutions labeled 1, 2, 3, 4, and 5 in separate disposable cups.
2. Get 3 cotton swabs and a cup of water for rinsing.
3. Dip a cotton swab into Solution 1 and touch it on your tongue until you taste it. If you don't taste it immediately, try the swab on a different area of the tongue.

❋ **If you need to get more of the same solution, use the clean end of the cotton swab. Do not double dip!**

4. Record the taste of Solution 1 in Table 16.1 below.
5. Rinse your mouth thoroughly with water.
6. Repeat steps 3–5 for Solutions 2, 3, 4, and 5 and record the results in Table 16.1. Be sure to rinse your mouth after each tasting.

❋ **If you are sensitive to MSG, do not taste Solution 5.**

Table 16.1
Results of Tasting Solutions

Solution	Taste
1	
2	
3	
4	
5	

Check with your classmates. Did everyone identify the solutions with the same descriptions?

The taste of Solution 5 has been recognized in Japan for nearly a century. Some Western researchers have recently concluded that it is a fifth basic taste, though not all agree. It is called umami (ooh mah me), a Japanese word without an exact English equivalent. *Savory, essence, pungent,* and *meaty* have been suggested to describe this taste. The umami taste is elicited by glutamate, an amino acid, as well as by some other chemicals. The solution you tasted was monosodium glutamate (MSG), which is used as a flavor enhancer. Glutamate is the most abundant amino acid and is present in many protein-containing foods including meat, seafood, and aged cheeses such as Parmesan. We can't taste it when it is bound with other amino acids in protein molecules, but cooking or other processing such as fermentation or aging releases the free amino acids.

A sweet taste can be caused by numerous chemicals besides sugar. Much effort has been devoted to the search for a low-calorie substitute for sugar. Aspartame (brand name NutraSweet) is a combination of amino acids that has been very successful as a sweetener. Some sweet-tasting substances are even poisonous, including some lead compounds, glycols (for example, antifreeze), and chloroform. However, those are not substances one would naturally encounter as food.

The salty taste is caused by various ionized salts. Table salt (NaCl), which produces sodium and chloride ions when dissolved in water, is the most familiar. The taste can also be elicited by related salts such as $CaCl_2$ (calcium chloride), KCl (potassium chloride), and NaI (sodium iodide).

Are sweet, salty, and umami tastes ones that you enjoy?

Why do you suppose our taste buds can sense sweet, salty, and umami tastes? (Hint: In what foods do these tastes occur?)

Sourness is really a measure of pH. The stronger the acid, the more sour the taste. A 5% solution of acetic acid (the same as vinegar) was used for the sour taste in this activity.

Is the sour taste one you enjoy or one you avoid? Why might it be adaptive to avoid food or drink that tastes sour?

Like the sweet taste, the bitter taste can be caused by various chemicals, especially alkaloids. Alkaloids are complex molecules that occur naturally in many plants. Some have powerful physiological effects, including death, in humans. Some have been used or modified for their medicinal properties. Caffeine, strychnine, nicotine, cocaine, morphine, and LSD are alkaloids. So is quinine, which was used as the tasting solution. Quinine is used in the treatment of malaria and gives the bitter flavor to tonic water.

Is the bitter taste one you enjoy or one you avoid? Why might it be adaptive to avoid food or drink that tastes bitter?

Table 16.2 shows the minimum concentrations of different substances needed to stimulate the sweet, sour, salty, and bitter taste receptors. (Information on the umami taste is not available.) Explain why these differences make sense in terms of human survival.

Table 16.2
Minimum Concentrations for
Taste Receptor Stimulation

Sucrose (sweet)	0.01M
NaCl (salty)	0.01M
HCl (sour)	0.0009M
Quinine (bitter)	0.000008M

Why is it important for animals, including humans, to have sense organs for taste in the mouth rather than elsewhere in the digestive system?

 Before you do the next activity, rinse your mouth out thoroughly with water.

Activity B: Individual Differences in Taste

Food preferences may be attributed to various factors such as texture that have nothing to do with taste, or to the number and sensitivity of a person's taste buds. In this exercise, you will investigate individual differences in the ability to taste certain chemicals.

Procedure

1. Get one piece each of the four color-coded taste papers: control (which has no chemical on it), PTC (phenylthiocarbamide), thiourea, and sodium benzoate.

2. Taste the control paper first by placing it on your tongue and allowing your saliva to dissolve any chemicals present in the paper. Record the taste in Table 16.3. Discard the control taste paper.

3. Taste each of the remaining papers in the order listed in Table 16.3 and record what you taste.

4. Discard each paper in the trash after you have tasted it.

Table 16.3
Results of Testing Taste Papers

Paper	Color	Taste
Control	White	
PTC	Blue	
Thiourea	Yellow	
Sodium benzoate	Pink	

Compare your responses to the taste papers with the responses of your classmates. Did everyone experience the same tastes? Explain any differences.

The ability to taste the chemical PTC is determined by a single gene. Tasters have at least one copy of the dominant allele and nontasters are homozygous recessive. In the natural world, would there be an advantage or disadvantage to being a "taster"?

Activity C: Smell Discrimination

In this activity you will explore your ability to identify odors. You should have several opaque containers that are covered tightly with lids. Each one is numbered so that you can record your observations, but it is not necessary to take them in order. Work with a lab partner. One person should be the subject and the other the experimenter. When you have finished, switch roles and repeat the procedure.

Procedure

1. The subject should close her or his eyes. Without looking into the container (because he or she will be the subject later), the experimenter should raise the lid of a container and hold it up to the subject's nose.

✳ **Be sure to close the lid after each trial to keep the materials fresh.**

---

2. The subject should identify the odor. Without telling your lab partner what you smelled, write this observation in Table 16.4 (you may open your eyes to do this!).
3. Repeat the procedure until the subject has identified the odor in each container.
4. Switch roles and repeat the procedure.

Table 16.4
Results of Smell Discrimination Testing

Jar	Smell
1	
2	
3	
4	
5	
6	
7	

Compare your results to your lab partner's results. Did you agree on identifying the odors? Did other students agree with your results? Explain.

In contrast to the four or five basic tastes we can identify, we can distinguish perhaps 10,000 odors. Scientists do not understand exactly how this is accomplished. Some believe that smells, like tastes, can be classified in several basic categories, but there is no agreement as to what those categories are. One proposed scheme groups smells as fragrant, burnt, "goaty," or acid. Another scheme groups them as fruity, spicy, flowery, foul, or resinous.

Activity D: Influence of Smell on Taste

✳ This test should be done after the other taste tests.

Most foods are complex mixtures of chemicals that stimulate our sensory receptors with the sweet, salty, bitter, sour, or umami taste, or some combination of them. In addition, we can perceive texture and temperature. Since the mouth also contains air, scent molecules from food enter the nasal cavity and stimulate the olfactory epithelium. Thus the sense of smell plays a large role in what we "taste."

Procedure

1. Close your eyes while your lab partner gets a Life Saver.
2. With your eyes still closed, pinch your nostrils closed while your partner puts the candy on your tongue.
3. Keep holding your nose while the candy dissolves in your mouth. Guess its flavor:

4. Release your nose. Does the Life Saver taste different? Guess its flavor again.

5. What is the actual flavor of the candy?

Did you think the flavor was the same when you could taste and smell as you did when you could only taste? Explain your results.

A similar demonstration can be done using any foods that have the same texture. For example, if you chew on a piece of raw onion while you hold your nose, you'll find its taste indistinguishable from the taste of raw potato.

EXERCISE 16.2
Vision

Objectives

After completing this exercise, you should be able to

1. Identify the lens, iris, pupil, retina, sclera, cornea, fovea, and choroid coat and give the function of each.
2. Describe the functions of the rods and cones.
3. Explain how eyes become adapted to the dark and to bright light.
4. Explain the physical basis of red-green color blindness.
5. Explain the presence of a blind spot.
6. Explain accommodation.

Vision is a result of our ability to sense and interpret light energy. When photoreceptors are stimulated by light, they transmit an action potential, which is carried to the brain. In this exercise you will learn about the properties of the photoreceptors, color vision, and some structural features of the eye that affect vision.

Activity A: Structural Features of the Eye

The eye is constructed so that images (patterns of light) will be focused on light-sensitive photoreceptors in the retina. A camera has the same function: The photographer wishes to focus an image on photosensitive film. A comparison of the eye and the camera (Table 16.5) will help you understand the functions of the various parts of the eye.

In this exercise you will examine the external features of the human eye. You will then dissect a sheep eye, which is structurally very similar to the human eye, to observe the internal features. Figure 16.3 illustrates the structures you will see.

Procedure

1. Look at your lab partner's eye and identify the sclera, cornea, iris, and pupil.

The white of the eye is the **sclera,** a tough coat that extends around the eyeball.

Part of the front of the sclera, the **cornea,** is transparent to permit passage of light. The cornea can be seen best by looking at a side view of the eye. Notice that it bulges out slightly. The cornea plays a role in bending the light that enters the eye; it acts as a fixed lens. If the curvature of the cornea is too great, myopia, or nearsightedness, results.

Table 16.5
Comparison of the Eye and a Camera

Part of Eye	Part of Camera	Function
Lens	Lens	Focuses light
Iris	Diaphragm	Regulates size of light opening
Pupil	Light opening	Allows light to enter
Retina	Film	Senses light
Sclera	Camera body	Protects the eye and helps maintain shape of the eyeball
Cornea	Colorless filter	Covers and protects light opening; bends light
Choroid coat	Black paint on inside of camera	Absorbs extra light to prevent internally reflected light from blurring the image

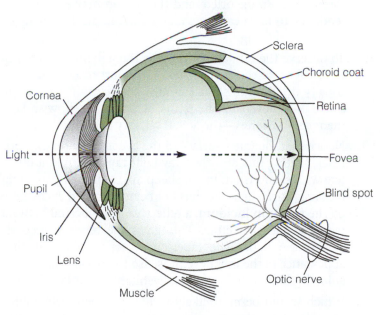

Figure 16.3.
The human eye.

The pigmented structure is the **iris.** It contains muscles to regulate the size of the **pupil,** which is simply an opening to allow light to enter.

2.   Get a sheep eye in a dissecting pan and a scalpel. You may wear latex gloves while you perform this procedure.

3.   Examine the external tissues. Notice that there is a large amount of fat. This cushions the eye in its socket.

4.   As you move aside the fat, look for the muscle tissue that held the eye in place. The muscles are attached to the sclera.

> ⚠ Handle the scalpel with caution. When cutting, be sure to cut away from your body and keep your fingers out of the way. The scalpel blade is extremely sharp.

5. Cut away the fat and muscle as best you can. You will see a thick white cord emerging from the back of the eye. This is the **optic nerve,** which transmits information from the sensory cells of the eye to the brain.

6. Locate the pupil, iris, sclera, and cornea of the sheep eye. The preservative caused the cornea to become cloudy; the cornea of any living eye is clear.

7. Place the eye in the dissecting pan with the cornea down. Make a circular incision about 5 mm ($^1/_4$ inch) above the cornea all the way around the eye. The small cavity between the cornea and the iris contains a watery liquid that helps maintain the shape of the eye. This liquid will drain out as you cut the eye.

8. Locate the lens, which is a hard white disk. Like the cornea, it is transparent in the living eye. A **cataract** is a clouding of the lens that prevents light from passing through it unobstructed. The image formed is then distorted, and the person may even become blind. A tendency to form cataracts may be inherited, but many cataracts are the result of aging.

9. To observe the retina, first remove the jellylike material that fills the large cavity of the eyeball. In the preserved eye, this substance has usually dried up to some extent. Its function is to help maintain the shape of the eyeball. The retina, which is a yellowish layer, contains the rods and cones—the actual sensory cells of the eye.

10. Notice that the inner surface of the eye is lined by the choroid coat, a black coating that absorbs stray light rays that might otherwise interfere with clear vision. In the sheep eye, as in many mammals (but not humans), stray light that has not been absorbed in the retina bounces off the **tapetum lucidum,** a reflective layer located between the choroid coat and the retina. This gives the retina a second chance to absorb the light. The tapetum is responsible for the iridescent appearance of the inner eye. It is also the reason that the eyes of some animals glow in the beam of a flashlight or car headlights.

    Which would be more likely to have a tapetum lucidum, nocturnal animals (those that are active primarily at night) or diurnal animals (those that are active during the day)? Explain your answer.

11. Locate the optic nerve again and follow it to its origin inside the eye. The gathering of neurons in this place results in a blind spot where no image is formed on the retina. The blind spot will be studied in Activity C.

Activity B: Photoreceptors

Light entering the eye falls on the retina, which contains neurons that are modified into photosensitive cells, the **rods** and the **cones.** The human eye contains about 125 million rods and 7 million cones.

The rods and cones contain pigments that absorb light energy, causing rapid (within nanoseconds) breakdown of the pigments. This change stimulates the nerve fiber leading to the brain. The final product of the breakdown is a form of vitamin A, which can be stored in the retina and later reconverted to photosensitive pigments.

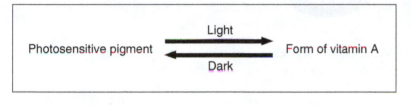

Figure 16.4.
Forms of photosensitive pigments. The pigments undergo a molecular change when they absorb light. After a short period of darkness, the change is reversed.

Night blindness is a condition in which people see well in the daytime but very poorly at night. Using Figure 16.4, suggest a mechanism and a cure for this affliction.

If your retinas are exposed to strong light over a long period of time, most of the photosensitive form of the pigment is converted to vitamin A and your eyes become less sensitive to light. If you then walk into a dark theater, for example, your eyes will not be sensitive enough to distinguish the light spots from the dark areas. In photographic terms, your image is underexposed—everything looks dark. In the dark, however, the photosensitive pigments are reconstructed from vitamin A, and within 1 minute sensitivity to light increases tenfold. After 40 minutes sensitivity increases 25,000 times. This is called **dark adaptation.**

When you walk out of the theater into the bright sunlight, the image you see is overexposed. The dark areas of the image look bright too, because your retina is sensitive even to dim light. Adaptation to light, though, is more rapid than dark adaptation, and will be complete in a few minutes.

You can demonstrate "local" adaptation of the retina by fixing your eyes on the white dot in the center of the peace symbol in Figure 16.5 for 30 seconds. Then look at the black dot at the right of Figure 16.5.

Figure 16.5.
Adaptation demonstration.

What do you see when you stare at the black dot?

Explain why.

Why don't you see an afterimage like this when you are reading black print on a white page?

Besides adaptation of the photoreceptors, what other mechanism helps the eye adjust to changes in light and dark? (Hint: Review Table 16.5.)

Usually during a cave tour the guide will turn off all lanterns and flashlights to let everyone experience total darkness. No matter how long you wait, you will not be able to see anything. Why not? Why don't your eyes adapt to the dark?

Cones are responsible for bright-light vision and for the perception of detail and color. Cones are concentrated in the center of the retina in an area called the **fovea,** which is directly in line with the center of the cornea and lens. The human fovea has approximately 150,000 cones per square millimeter, but some birds have up to a million cones in the same area.

There are three types of cones, called blue, green, and red, which contain different photosensitive pigments. The red cones are sometimes called yellow because the photopigment they contain also absorbs yellow light. Figure 16.6 shows the range of wavelengths of light absorbed by each type of cone.

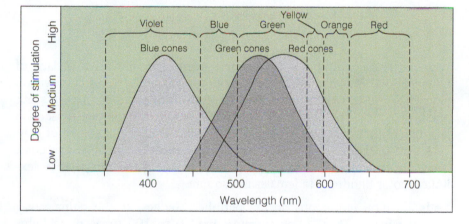

Figure 16.6.
Cone sensitivity to light. Each type of cone contains a different photopigment that absorbs light. When light is absorbed, the cone is stimulated to produce an action potential.

When a photopigment absorbs light, the cone is stimulated and sends an action potential to the brain. As Figure 16.6 shows, the blue cones are maximally stimulated in response to light of 420 nm. The blue cones can also absorb light at other wavelengths. For example, when light of 530 nm is received by the retina, green cones are stimulated maximally. But both the blue cones and the red cones are also stimulated to some degree. All three curves overlap.

We see white when all the cones are stimulated and black when none of them is stimulated.

The reason that we see so many gradations of colors is the overlap in the curves. The brain interprets the *ratio* of stimulation of the three cones. For example, light with wavelengths from 580 to 595 nm stimulates both the red and the green cones, and the brain interprets that color as yellow. This can be demonstrated by using red and green colored filters to make red and green light. The red light stimulates the red cones and the green light stimulates the green cones.

Procedure

1. Aim two light sources so that the beams overlap.
2. Hold a red filter in front of one light source and a green filter in front of the other.
3. Position the lights so that all three colors (red, green, and a third color where red and green intersect) can be seen, and project the beams onto a sheet of white paper.

What color do you see in the area where the lights overlap?

Two percent of all men lack red cones. Using Figure 16.6, estimate which colors these people have the most difficulty seeing. Explain your answer.

Six percent of all men lack green cones. Using Figure 16.6, estimate which colors these people have the most difficulty seeing. Explain your answer.

Both of these conditions are called red-green color blindness. A lack of blue cones is also possible, but it is quite rare. Other types of color-vision deficiencies result when individuals have all three types of pigments but one is defective. The only truly color-*blind* people, though, are those who lack two of the three types of cones; they see everything in shades of gray. Less than 0.01% of the population has this affliction. Because the color genes are located on the X chromosome, meaning that the trait is sex-linked, color blindness in females is also rare.

Rods are more numerous on the periphery (edges) of the retina, away from the fovea. These photoreceptors are 50 to 100 times more light-sensitive than the cones. They function in dim-light vision and are insensitive to colors.

Some animals are so specialized for diurnal or nocturnal existence that their retinas contain only one type of photoreceptor.

Gray squirrels and some birds are purely diurnal. Their retinas contain only _____.

Bats and owls, which are strictly nocturnal, have only _____.

If you are trying to see an object in dim light, should you look at the object directly so that the image falls on the cones or should you look at it out of the side of your eye so that the image falls on the rods? Explain your answer.

Activity C: The Blind Spot

The retina is supplied with blood vessels and neurons, which are in front of the photoreceptors. As you saw when you dissected the sheep eye, the neurons all converge at one point, called the optic disk, where the optic nerve exits the eye and carries impulses to the brain. There are no photoreceptors at the optic disk, so any light that falls on this area cannot be perceived; there is a **blind spot.** You can demonstrate the presence of the blind spot with the following procedure.

Procedure

1. Hold up your lab manual with Figure 16.7 about 12 cm (5 inches) in front of you.

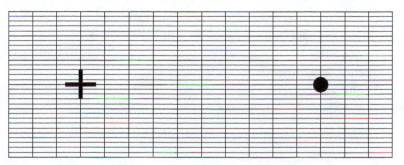

Figure 16.7.
Demonstration of blind spot.

2. Close your left eye and focus the right eye on the plus sign in the figure.
3. Very slowly move the lab manual away from you until the circle disappears.
4. Repeat the procedure with the right eye closed: Focus on the circle with the left eye and move the book until the plus sign disappears.

When you locate your blind spot, the circle or the plus sign disappears, but you still see something. What do you see?

Even though you have a blind spot in each eye where there are no rods or cones, you don't perceive a blind spot in your vision under normal circumstances. Why not? (Hint: You usually use both eyes to look at an object.)

Activity D: Accommodation

Muscles holding the lens change the shape of the lens in order to focus on objects at different distances. The ability to focus on a close object, which requires the lens to become almost spherical, is called **accommodation.** During the aging process, physical changes in the lens itself affect its ability to change shape. A 10-year-old can focus on a point 9 cm from the eye, but a loss of accommodation is usually noticeable by age 40 and typically must be corrected with reading glasses when a person reaches the mid-40s. In the following procedure, you will measure your accommodation distance.

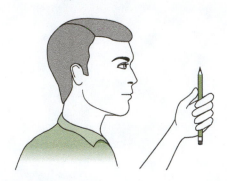

Figure 16.8.
Measuring accommodation distance.

Procedure

1. Cover one eye and hold out a pencil at arm's length.

2. Focus on the point and slowly bring the pencil in toward the eye until it is no longer in focus, as shown in Figure 16.8.

3. Have your lab partner measure the distance between the eye and the pencil. Record your measurements.

   Right eye:_____cm.

   Left eye:_____cm.

   If you wear glasses, repeat the procedure without your glasses.

   Right eye:_____cm.

   Left eye:_____cm.

Your instructor will collect data from everyone in the class. Explain any similarities and differences you see among the members of the class.

E X E R C I S E   1 6 . 3

Skin Senses

Objectives

After completing this exercise, you should be able to

1. Explain adaptation and distinguish between rapidly and slowly adapting receptors.

2. Discuss how receptor adaptation affects our perception of temperature.

We receive a great deal of information about the external environment through sensory receptors localized in different places in the skin. A variety of receptors contribute to your tactile sensitivity, or sense of touch. There are also thermoreceptors that sense temperature.

Activity A: Sensitivity to Pressure

One measure of sensitivity of touch is the **two-point threshold,** the minimum distance that must be between two points before they are felt as separate points. In a very sensitive area, this distance may be as little as 1 mm. That is, if you are touched by two points that are 1 mm apart, you feel two touches. On a less sensitive part of the body, two points 1 mm apart will feel like one touch.

Procedure

1. Get a compass and a ruler.
2. Using the ruler to measure point-to-point distance, set the points of the compass 5 mm apart.
3. While your lab partner has his or her eyes closed, touch both points simultaneously to the palm of his or her hand.

 Be careful not to stimulate pain receptors!

4. If your partner feels two separate points, reduce the distance between the points. If he or she feels only one point, increase the distance. Your goal is to find the smallest distance at which your partner can still distinguish two separate touches. Record this distance in Table 16.6.
5. Repeat the procedure for the other parts of the body listed in Table 16.6.

Table 16.6
Touch Discrimination

Location	Two-Point Threshold (mm)
Palm	
Fingertip	
Lips	
Outer forearm	
Outer shoulder	
Calf	

 While experimenting, your point-to-point distance should range between 1 mm and 30 mm. Touch your partner with only one point from time to time as a test.

Why might certain points of your body have lower two-point thresholds than other parts?

Compare your results with those of your classmates. Are there differences among individuals for the two-point threshold?

Activity B: Perception of Temperature

There are at least four types of nerve endings involved in sensing temperature. Figure 16.9 shows the temperature range that stimulates each type. Note there is some overlap between cold and warmth receptors. As is the case with the cones and color vision, our perception of temperature results from the relative degrees of stimulation of each type of receptor. Imagine that you get into a tub filled with water that is 35°C (95°F, slightly below body temperature). Both cold and warmth receptors are stimulated and we perceive this combination as lukewarm. On the other hand, if the water is 25°C (77°F) only cold receptors are stimulated, and we definitely perceive cold.

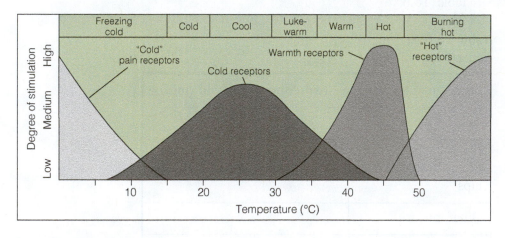

Figure 16.9.
Response of thermal receptors.

Our perception of temperature is complicated by the fact that most areas of the body have three to four times as many cold receptors as warmth receptors. The number of receptors per body part also varies, from 15 to 25 cold receptors per square centimeter in the lips to fewer than 1 cold receptor per square centimeter in broad surface areas such as the back.

Certain chemicals can also stimulate the temperature receptors. Mint can give a sensation of cold, while mustard can give a feeling of warmth. An old folk remedy for colds was to use a mustard plaster on the chest. The sensation of heat from rubbing compounds such as Ben-Gay is also a result of chemical stimulation of temperature receptors.

Nearly all receptors have the ability to **adapt;** that is, they stop responding to the stimulus. When a receptor has adapted completely, it no longer sends a message to the central nervous system that a stimulus is occurring. (We have already discussed the adaptation of photoreceptors to light.) Some tactile receptors adapt rapidly. They react while a change is taking place, so they transmit a signal when light pressure begins and another one when the pressure stops. In this way they detect movement. You are usually not aware of the clothes on your body because the touch receptors on the skin have adapted. However, if someone tugs at your sleeve, you feel the movement of the cloth against your skin.

Other receptors adapt slowly: The receptor continues to transmit information to the brain as long as the stimulus is present. Tight-fitting clothes stimulate slow-adapting pressure receptors that make you uncomfortably aware that something is constricting your body.

The nerve endings that are specialized as pain receptors do not adapt (or adapt very little). Why?

The thermal receptors, on the other hand, can adapt to a great extent, as the following procedure shows.

Procedure

1. Working with a lab partner, prepare three beakers of water, with temperatures approximately 20°C, 33°C, and 44°C. Label them A, B, and C, respectively.

2. Dip a different finger briefly into each beaker and record your perception of the temperature (hot, indifferent, etc.).

    A:_____   B:_____   C:_____

3. Measure the temperature in each beaker and record the actual temperatures.

    A:_____   B: _____ C: _____

4. Immerse the index finger of your right hand in Beaker C and your left index finger in Beaker A. Record the time it takes for each finger to adapt.
    Right:_____ Left:_____

 This will take several minutes. Don't abandon your lab partner! Examine Figure 16.9 and make predictions about the outcome of the experiment while you wait.

5. After adaptation has occurred, move both fingers to Beaker B. Record how each finger feels.
    Right:_____ Left:_____

Explain what has happened in terms of receptor adaptation: What happens when the cold receptors adapt? What happens when the warmth receptors adapt? Why do your fingers feel different when they are moved to the second beaker?

As a consequence of adaptation, the thermal receptors respond noticeably to changes in temperature. When the cold finger in Beaker A moved to water that was warmer than the finger, the temperature of the finger began rising rapidly. This change in temperature made your finger feel much warmer than it would if it were already close to the temperature of the water. Thus, our experience of warmth or cold is strongly influenced by changes in temperature. List a few examples of situations when a change in temperature exaggerates the sensation of warmth or cold:

Your ability to perceive a rapid temperature change is proportional to the area of skin involved. If your whole body is subjected to a change as small as 0.01°C, you may be able to detect it. If a very small area of the body is involved, you might not notice a temperature change 100 times that great (1°C).

Considering all you have learned about perception of temperatures, would you advocate the cautious slow-wading/splashing approach to getting wet in the cold ocean or the quick dunk? Defend your choice.

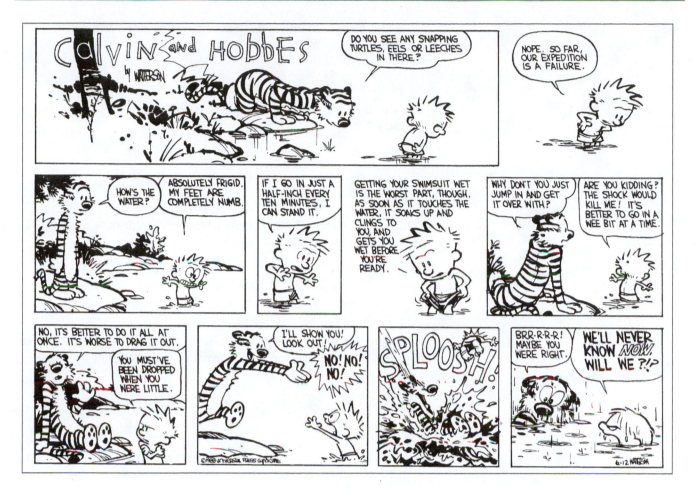

Questions for Review

1. Explain how information about the environment gets to the brain.

2. If you close your eyes and pinch your nostrils closed, is it easier for you to distinguish between lime Kool-Aid and lime juice or between lime Kool-Aid and cherry Kool-Aid? Explain your answer.

3. Explain how the substances that stimulate taste and smell receptors are similar to each other.

4. Dogs and horses are only able to see images in black and white. Give a reason why this might be the case.

5. Explain the scientific basis for the saying "All cats are gray in the dark."

6. Which photoreceptors are primarily responsible for dark adaptation? Explain how dark adaptation works.

7. How does the retina of a person who is red-green color-blind differ from the retina of a person with normal color vision?

8. Explain why older people complain that their arms seem to be getting shorter when they pick up something to read.

9. Compare two-point touch discrimination with the resolving power of the microscope. (Recall from Lab Topic 4 that resolving power is the ability to distinguish two separate points.)

10. How does the ability of tactile receptors to adapt slowly or rapidly affect the kind of stimuli they sense?

11. You've been working outside on a cold day and when you come inside you wash your hands. Even though the water is lukewarm, you feel like your hands are burning. What happened?

Plant Structure

Introduction

As discussed in Lab Topic 13 (Animal Diversity), all birds have structural similarities that allow us to identify certain animals as birds, whether they are owls or ostriches. Although different species of birds may be quite diverse in appearance and behavior, they also have many structural features in common, including feathers, wings, scaly feet, and horny beaks. Similarly, despite the diversity of adaptations in the approximately 280,000 species of plants, most plants are recognizable as plants because of certain features that they share. Stems, leaves, and roots are included in our concept of "plant," and indeed almost all plants have these structures. In addition, almost all plants contain the green pigment chlorophyll, which captures the sun's energy for photosynthesis. All of these characteristics enable us to look at an organism and immediately classify it as a plant.

In this lab topic you will study plant structure by examining the cells and tissues of several different angiosperms, or flowering plants. As you may recall from Lab Topic 12 (Plant Diversity), angiosperms are vascular plants. That is, they contain tissues that carry food and water throughout the plant. Gymnosperms (for example, pines) are also vascular plants, as are ferns and their relatives. Lower plants such as mosses lack vascular tissue, which restricts them to moist habitats. However, plant cells are similar in all types of plants, and because of the presence of chlorophyll, we have no trouble recognizing mosses and liverworts as members of the plant kingdom along with flowering plants.

Outline

Exercise 17.1: Plant Cells and Tissue Systems

Exercise 17.2: Plant Organs

    Activity A: Leaves

    Activity B: Roots

    Activity C: Stems

Exercise 17.3: Dermal Tissue

    Activity A: Dermal Tissue in Stems

    Activity B: Dermal Tissue in Roots

    Activity C: Dermal Tissue in Leaves

Exercise 17.4: Ground Tissue

    Activity A: Parenchyma

    Activity B: Collenchyma

    Activity C: Sclerenchyma

E X E R C I S E   1 7 . 1
Plant Cells and Tissue Systems

Objectives

After completing this exercise, you should be able to

1. Describe the characteristics of primary and secondary cell walls.
2. Define the three tissue systems.

In this exercise you will review some features of plant cells that distinguish one cell type from another. A living cell has a protoplast. The protoplast consists of the cytoplasm, or cell contents, and the nucleus. The cytoplasm contains the enzymes and organelles that carry out the metabolic processes of the cell. The nucleus contains the genetic material. The protoplast of a plant cell is bounded by the cell membrane, as it is in animal cells. In addition, plant cells have a cell wall outside the membrane. As you may recall from Lab Topic 4 (Using the Microscope), plant cells contain vacuoles that take up water. The cell wall, though permeable to water, serves to limit expansion of the protoplast as the vacuole swells.

Living cells, those that are carrying on metabolic processes or are actively dividing, have primary cell walls. Primary cell walls are composed mostly of cellulose, a polymer made up of glucose subunits, and are somewhat stretchable to accommodate growth. As you will see in this lab topic, primary cell walls may be thin and delicate or they may be thick and tough.

Certain cells also have a **secondary cell wall,** which is laid down after the cell matures. The protoplast often dies after depositing the secondary wall, so no further metabolic activities take place in the mature cell. Secondary walls are strong and rigid and are mostly found in specialized cells that have a support function. Another polymer, lignin, preserves the secondary cell walls of many cells as well as strengthening them. We take advantage of the strength and longevity of secondary cell walls when we use wood, which consists entirely of lignified secondary cell walls.

Besides the cell wall, plant cells are also distinguished from animal cells by the presence of chloroplasts. Chloroplasts are the site of photosynthesis and are one of the organelles located in the cytoplasm. Since they contain the green pigment chlorophyll, they are readily apparent by light microscopy of unstained materials.

Compared to the bodies of animals such as humans, plant bodies are marvels of simplicity. The organs are few and are relatively similar throughout the higher plants: roots, stems, leaves, and reproductive structures, which are specialized leaves. Plant bodies are all made up of the same cell and tissue types, which are very similar throughout the plant. Because of this continuity, three tissue systems can be identified in the plant body (Figure 17.1).

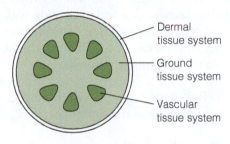

Figure 17.1.
The three tissue systems in a typical dicot stem.

In a simple dicot stem the dermal tissue system consists of the epidermis. In the plants you will examine for this lab topic, the epidermis is a single layer of cells.

The vascular tissue system consists of the xylem (water-conducting) and phloem (food-conducting) tissues. In stems, leaves, and some roots the vascular tissue is found in distinct clusters called vascular bundles.

Vascular tissues are arranged differently in the roots, stems, and leaves. The arrangement also differs between monocotyledons (one seed leaf, monocots) and dicotyledons (two seed leaves, dicots), which are the two large subdivisions of flowering plants. The basic organs, tissues, and cells are the same in both monocots and dicots. The examples used in this laboratory will all be dicots. Other characteristics of the two subdivisions will be studied in Lab Topic 18 (Flowers, Fruits, and Seeds).

The ground (fundamental) tissue system consists of everything that isn't vascular tissue or dermal tissue. Parenchyma, sclerenchyma, and collenchyma cells make up the ground tissue. You will learn the characteristics of these cells in this laboratory.

E X E R C I S E   1 7 . 2

Plant Organs

Objectives

After completing this exercise, you should be able to

1. Describe the primary functions of leaves, roots, and stems.
2. Briefly describe how the microscope slides that you viewed were prepared.

Activity A: Leaves

The primary function of leaves is photosynthesis. Their structure is designed to provide surface area to intercept sunlight. The leaf surface also has openings, called **stomata**, to permit entry of the CO_2 that is needed for photosynthesis. Internally, the cells and tissues of leaves are either involved in carrying out photosynthesis, in bringing water and minerals to the cells that carry out photosynthesis, or in exporting the sugars produced by photosynthesis to other parts of the plant.

In this procedure you will begin your examination of a prepared slide of a thin section of a leaf. The slide was made using a **microtome**, an instrument that slices tissue so thinly that light can pass through it. You may recall from Lab Topic 4 (Using the Microscope) that tissues are also stained so that various structures can be seen. In the prepared slides that you will view for this lab topic, primary cell walls are stained green and secondary cell walls are stained red. In most cells you will not be able to distinguish any features of the protoplast.

Procedure

1. Get a prepared slide of a typical dicot leaf cross section.
2. Place the slide on the stage of a compound microscope and secure it with the stage clips.
3. Using the low-power objective (10×), focus on the center of the leaf.
4. Use one of the blank pages at the end of this topic to make a sketch of the leaf cross section. For this sketch, simply draw an outline of the leaf. You will add details later as you learn about the cells and tissues.

Activity B: Roots

Roots serve to anchor the plant in the soil, where they absorb water and minerals that are then conducted to the rest of the plant. Many plants also store food in the root until the next growing season. We take advantage of the stored sugar or starch in roots such as carrots, radishes, sweet potatoes, and turnips.

In this procedure you will begin your examination of a prepared slide of a thin section of a root. The root section was made using the same procedure that was used for the leaf section.

Procedure

1. Get a prepared slide of a typical dicot root cross section.
2. Place the slide on the compound microscope and focus on the root section using the low-power (10×) objective.
3. Sketch an outline of the root cross section following your sketch of the leaf. You will add details later as you learn about the cells and tissues of the root.

Where in the root are the cells that have secondary walls located?

Activity C: Stems

Stems support the leaves, enabling them to be positioned advantageously for photosynthesis. They also serve as a conduit for the exchange of materials between the leaves and roots. Some stems, for example cactus stems, are able to photosynthesize. Stems may also be a significant food storage site, as in asparagus and Irish (white) potatoes.

Instead of using a prepared slide, you will make your own cross sections of a *Coleus* stem. Your instructor has prepared short stem pieces that are supported by paraffin to make slicing easier. A crude microtome was made by a nut-and-bolt assembly. It will help you make the very thin slices that are desirable for viewing on the compound microscope. After obtaining thin sections, you will apply a stain called toluidine blue to help you distinguish the different cell types.

Procedure

Before beginning, read through all of the instructions and set up the three dishes for staining as directed. Be careful with the toluidine blue, since it can stain your skin and clothing.

Cutting the Sections

1. Get a nut-and-bolt microtome from your instructor. It contains a short piece of *Coleus* stem embedded in paraffin.
2. Hold the head of the bolt flat on the table with one hand (your left, if you are right-handed). Holding the razor blade in your other hand, remove the excess wax on top by slicing it down even with the top of the nut (Figure 17.2).

Slice as illustrated in Figure 17.2 to keep your fingers out of the way of the razor blade.

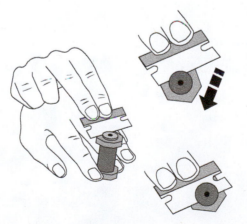

Figure 17.2.
Using the microtome.

3. Twist the bolt just a little so that a wafer-thin section sticks up out of the nut.

4. Use a slicing motion to cut this section down to the nut. When slicing, use as much of the razor blade as possible. That is, start on one end of the blade and slide down to the other with each slice, as shown in Figure 17.2. This technique results in a cleaner cut than you would get using a chopping motion.

5. As each slice is cut, transfer it to a small dish of 50% ethanol.

6. Continue to twist up thin sections of the block and slice them off, trying to get the thinnest possible slice each time. It is better to get a crescent of a thin slice than a thick slice that is entirely round. A paraffin block of this size will yield several sections. The last few millimeters of the block will fall out of the nut and can't be sliced.

Staining

1. Leave the stem sections in 50% ethanol for 5 minutes. Free the plant tissue from the paraffin surrounding it if necessary.

2. Using forceps, move the sections to a dish of toluidine blue stain and leave them there for 5–10 minutes.

3. Rinse the sections in a dish of distilled water.

4. Mount each section in a drop of 50% glycerine on a slide. You can put two or three sections on each slide. Mount all of your sections so you can find the best one.

5. Add a coverslip and observe your sections on the compound microscope.

6. Following your leaf and root sketches (Activities A and B), draw an outline of a wedge of your best section. You will add details later as you learn about the cells and tissues of the stem.

In general, toluidine blue stains secondary cell walls green to blue-green or blue. Primary cell walls stain red or purplish colors.

Where are the cells with secondary walls located in the stem?

The cross sections of the stem and the leaf are clearly different shapes. Considering the function of each organ, why might their forms be different?

E X E R C I S E   1 7 . 3
Dermal Tissue

Objectives

After completing this exercise, you should be able to

1. Describe the general functions of dermal tissue.
2. Identify the dermal tissue on a slide or diagram of a stem, root, and leaf cross section.
3. Explain how and why the epidermis of roots and stems differs.
4. Identify guard cells and stomata and explain their function.

The **epidermis** is the outermost layer of cells; it serves as a covering for the plant body. As the plant's interface with the environment, the epidermis regulates the passage of substances to and from the environment. Epidermal cells have primary cell walls and retain their protoplasts. For the above-ground plant parts, the epidermis provides protection against water loss, which is a major problem for land-dwelling plants. Some epidermal cells deposit a coating called a **cuticle** to cover the outer wall. The cuticle consists of lipid substances that are impermeable to water, which helps the plant retain water. The cuticle is especially important to plant parts such as fruits that have a high water content and to plants that inhabit very dry regions.

Activity A: Dermal Tissue in Stems

In this procedure you will examine the dermal tissue system of the *Coleus* stem. As an above-ground organ, the stem must be protected against water loss. This is the primary function of the stem epidermis. It is typically only a single layer of cells, which may be covered with a cuticle. Many stems and leaves also have specialized epidermal outgrowths such as hairs that, in some plants, aid in preventing water loss or discourage predation by insects. You should see abundant hairs on your *Coleus* stem section.

Procedure

1. Identify the epidermis in your *Coleus* stem section.
2. On the stem outline that you sketched previously, draw and label several representative epidermal cells.

Activity B: Dermal Tissue in Roots

The epidermis of young, active roots is very permeable to water. It is the function of roots to absorb water and minerals into the plant. In these procedures you will examine the dermal tissue system of the root and root hairs in germinating seeds.

Procedure for Examining Root Dermal Tissue

1. Return your dicot root slide to the microscope and identify the epidermis.
2. On the root outline you sketched previously, draw and label several representative epidermal cells.

In addition to the epidermal cells that you labeled on your root cross section, part of the root has specialized epidermal cells with extensions called root hairs. Root hairs are found just behind the growing tip of a root, and they are responsible for most of the absorption of water and minerals into the plant. Since they are thin, the root hairs collectively have an extremely large surface area for absorption. Gardeners refer to these active, growing roots as feeder roots and take care not to damage the delicate hairs during transplanting.

Procedure for Observing Root Hairs in Germinating Seeds

1. Get a germinated seed from your instructor and place it on a clean microscope slide. No coverslip is needed.
2. Observe the root hairs using either a dissecting microscope or the scanning-power objective of a compound microscope.
3. Sketch the root hairs of the germinated seedling in the margin of your lab manual.

The lining of the small intestine in animals has many microscopic projections that create a vast surface area. What do you think is one role that the small intestine plays in the digestive system? Explain by relating that role to the large surface area.

Activity C: Dermal Tissue in Leaves

Because leaves have such a large surface area, they are particularly vulnerable to water loss. The epidermal cells of leaves fit together tightly and, as in stems, may derive additional protection from a cuticle. However, since the primary function of leaves is photosynthesis, the leaf must be at least partially open to the atmosphere in order to take in the carbon dioxide that is needed. Stomata (singular: stoma) are the openings that allow CO_2 to enter, but they do not remain open continually. Instead, they are regulated by a pair of guard cells that border the opening. In general, the stomata are open during daylight and when the CO_2 concentration in the leaf is low. They close in the dark and when the CO_2 concentration in the leaf is high. Stomata also close to conserve water if the plant is in danger of wilting.

In these procedures you will examine the dermal tissue system of the leaf by cross section and by making an epidermal peel.

Procedure for Examining Dermal Tissue Cross Section

1. Place the leaf cross section on the microscope.
2. Locate a stoma, which is a gap in the epidermis with one guard cell on each side of it. Stomata are more abundant on the lower surface of the leaf.
3. On your previous drawing of the leaf cross section, draw and label a pair of guard cells and several other epidermal cells.

Procedure for Examining Epidermal Tissue

Epidermal tissue can also be examined by peeling that layer of cells from a leaf. When you view your slide, you will be looking down on the cells, much like examining the squamous epithelium from your cheek in Lab Topic 4 (Using the Microscope), rather than viewing a cross section as you have done for the other slides viewed thus far in the lab.

1. Get a leaf from your instructor.
2. Put a drop of water on a microscope slide.
3. Fold the leaf over so that its lower surface is broken. Using forceps, grasp the edge of the epidermis that is exposed along the break and pull gently away from the break as shown in Figure 17.3. A thin sheet of tissue should peel off. If you're not successful, try again.
4. Float the epidermal peel on the drop of water on the slide. If your peel includes a thick section of tissue, cut that off with a razor blade before you add the coverslip.
5. Add a coverslip and view the tissue under the compound microscope. Draw several cells, including at least one pair of guard cells, in the margin of your lab manual.

How are the guard cells different from the other epidermal cells?

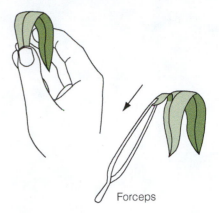

Forceps

Figure 17.3.
Making an epidermal peel.

EXERCISE 17.4
Ground Tissue

Objectives

After completing this exercise, you should be able to

1. Identify the areas of ground tissue on a slide or diagram of a stem, root, and leaf cross section.
2. Identify parenchyma cells on a slide or diagram of a stem, root, and leaf cross section.
3. Describe the general functions of parenchyma tissue in stems and leaves.
4. Describe the general function of collenchyma and sclerenchyma tissue.
5. Identify collenchyma cells on a slide or diagram of a stem and leaf cross section.
6. Identify sclerenchyma fibers on a slide or diagram of a stem cross section.
7. Identify a slide or diagram of sclereids.

Plant ground tissue consists of parenchyma, collenchyma, and sclerenchyma cells. Ground tissue serves more than one function. Depending on its composition and location, ground tissue may serve for storage, photosynthesis, structural support, or protection.

Activity A: Parenchyma

The most abundant plant cell type is parenchyma. These cells are relatively unspecialized and retain their protoplasts throughout their existence. They have primary cell walls and make up the mesophyll of leaves (where most photosynthetic activity takes place), the flesh of fruits, the pith of stems, and parts of the root and stem. In roots, stems, and fruits, many parenchyma cells are used to store food. Most of a sweet potato (root), white potato (stem), and the edible portions of a banana, for example, consist of parenchyma cells that are filled with starch grains.

Leaf Parenchyma

The ground tissue of the leaf, called the mesophyll, consists mainly of parenchyma cells. Your leaf cross section includes two distinct types of parenchyma in the mesophyll. The cells in the upper layer are elongated and loosely packed. Most of the photosynthesis in the leaf is carried out in these cells. The rest of the parenchyma consists of cells that are irregularly shaped and widely separated by spaces.

Procedure

1. Place the leaf cross section on your microscope.
2. Focus on an area of the mesophyll.
3. On your previous drawing of the leaf cross section, sketch and label several mesophyll cells. Include both types of parenchyma.

Why are the leaf parenchyma cells so loosely packed? (Hint: The spaces between the parenchyma cells are continuous with the stomata.)

Root Parenchyma

In the root cross section that you are studying, the ground tissue consists of the thin-walled parenchyma cells that surround the central axis. This area is called the cortex. The cells contain stored starch and are loosely packed to allow water and oxygen to pass through the spaces between the cells. Because they have primary cell walls, the parenchyma cells are stained green in this prepared slide.

Procedure

1. Place the root cross section on your microscope.
2. Focus on the cortex.
3. On your previous drawing of the root cross section, sketch and label several representative parenchyma cells.

Why do root cells require oxygen?

The innermost layer of cells in the root cortex is the endodermis. These cells fit tightly together and parts of their walls are impregnated with a lipid substance that is impermeable to water, much like the cuticle. At this point, water entering the plant is forced into the cells rather than being permitted to flow between them.

Why is there no cuticle on the root epidermis?

Stem Parenchyma

On your *Coleus* stem section, parenchyma is located in the center of the stem, which is called the pith. The cell walls are thin and, because they are primary cell walls, they should be stained a reddish-purple color by the toluidine blue.

Procedure

1. Place your *Coleus* stem section on your microscope.
2. Focus on the pith area (center).
3. On your previous drawing of the stem section, sketch and label several representative parenchyma cells.

Where else in the stem can you identify parenchyma cells?

What similarities can you observe between the parenchyma cells in the root cross section and in the stem cross section?

Activity B: Collenchyma

Many young stems and leaves contain collenchyma cells for support. These cells have primary cell walls and their protoplasts remain alive. An identifying characteristic of collenchyma is the noticeable thickening at the corners of the cell walls, which strengthens the wall.

Stem Collenchyma

Coleus, like other members of the mint family, has a square stem. You may have noticed this feature when you made the stem cross sections in Exercise 17.1. The stiffness of the stem, particularly at the four ribs, is due to collenchyma cells.

Procedure

1. Place the stem cross section on your microscope.
2. Focus on the layers of cells just below the epidermis. Look for the characteristic thickened corners that identify collenchyma cells.
3. On your previous drawing of the stem cross section, sketch and label several collenchyma cells.

Leaf Collenchyma

Collenchyma cells are also found as support tissue in leaves. When you look at a leaf, you will notice the veins that contain the vascular tissue. The midrib, or central vein, is especially prominent. In many leaves, including the one you are examining in this lab, collenchyma cells support the midrib. Collenchyma cells are also used for support in the petiole, which is the part of the leaf that connects the blade to the stem. Celery stalks are enlarged petioles, and the strings in celery are collenchyma cells.

Procedure

1. Place the leaf cross section on your microscope.
2. Locate a vascular bundle. Look for collenchyma cells below the vascular bundle.
3. On your previous drawing of the leaf cross section, sketch and label several collenchyma cells.

Activity C: Sclerenchyma

Like collenchyma, the function of sclerenchyma is to support and give strength to plant parts. However, sclerenchyma cells typically have hardened secondary cell walls; *skleros* is Greek for hard. Mature sclerenchyma cells may not have a living protoplast. There are two distinct varieties of sclerenchyma cells, fibers and sclereids. Your stem cross section may show fibers, but neither the root slide nor the leaf slide shows either type of sclerenchyma.

Fibers

Fibers are elongated sclerenchyma cells; some are as long as 25 cm (almost 10 inches). Fibers may be found in leaves, stems, and fruits. Cotton fibers grow on the surface of the seed, but fibers are more commonly associated with vascular tissue. Besides cotton, other plant fibers that are used to make textiles include flax, which is used to make linen, and ramie, which is sometimes blended with cotton or linen. Other plant fibers are used to make rope, twine, burlap, canvas, mattress stuffing, and many other products. The luffa, a fruit related to cucumber and squash, is another example of plant fibers. Once the fruit is dried and the fiber skeleton is exposed, it is used as a scratchy bath sponge.

Procedure

1. Place the *Coleus* stem cross section on your microscope.

2. Locate a vascular bundle. Not all stem sections have fibers, but if fibers are present, they will be found in a group just outside the vascular bundle. Fibers can be identified by their thick secondary cell walls and geometric shape.

3. On your previous drawing of the stem cross section, sketch and label a group of fiber cells.

Collenchyma cells and sclerenchyma fibers are similar in form and function. What are their differences, and why might these differences be important to a plant?

Sclereids

Sclereids are relatively short cells with very thick, typically lignified cell walls. Although they may be found in any part of the plant, you will most readily recognize them as components of fruits, nutshells, and other hard coverings such as the peach pit. Because they are so hard, they are sometimes called stone cells.

Pears contain clusters of sclereids that give them their characteristic gritty texture. In this procedure you will isolate pear sclereid cells and observe them microscopically.

Procedure

1. Get a bit of pear flesh on the end of a teasing needle and put it on one end of a microscope slide.

2. Use the dissecting scope to locate a cluster of sclereids. They are large, round, white structures.

3. With the teasing needle, push the sclereid cluster to the other end of the slide to separate it from the pear parenchyma.

4. Add a drop of water and a coverslip to the slide.

5. Apply gentle pressure to the coverslip to flatten the sclereids.

 It takes a fair amount of pressure to break open a cluster of sclereids, but be careful not to smash the coverslip.

6. Observe the sclereids on the compound microscope and, in the margin of your lab manual, sketch what you see.

Why do you think you found collenchyma and sclerenchyma cells in stem and leaf tissue, but not in the root?

EXERCISE 17.5
Vascular Tissue

Objectives

After completing this exercise, you should be able to

1. Describe the general function of xylem and list its cell types.
2. Identify xylem on a slide or diagram of a root, stem, and leaf cross section.
3. Describe the general function of phloem and list its cell types.
4. Identify phloem on a slide or diagram of a root, stem, and leaf cross section.

Xylem and phloem, the components of vascular tissue, are usually found together. Vascular bundles, discrete areas of vascular tissue separated by ground tissue, are a common arrangement in stems and leaves. Roots generally have a central vascular cylinder where the xylem and phloem are located.

Xylem is a tissue composed of different cell types that interact to conduct water and minerals through the plant. The cells that actually carry the water were misnamed "tracheary elements" in the seventeenth century, when it was thought that they served a function similar to that of the air ducts (tracheae) in animals; no one bothered to correct the name when the truth was discovered. The earliest tracheary elements to evolve in plants were similar to sclerenchyma fibers with thick, lignified secondary walls. These cells, the **tracheids,** are arranged with overlapping ends and a series of membrane-covered pits for passing water from one cell to the next. Angiosperms, which are the most highly evolved plants, have **vessel elements** in addition to tracheids. Since they are shorter, have larger diameters, and have holes in the cell walls that are not covered by membranes, vessel elements offer less resistance to water flow than do tracheids. Neither vessel elements nor tracheids contain protoplasts at maturity. Xylem tissue may also contain parenchyma and fibers.

Wood is xylem. We take advantage of the structural strength of its lignified secondary cell walls by using wood for building and making furniture. The softer woods, especially from conifers, are used to make paper.

Phloem is a tissue consisting of different cell types that interact to distribute the food made in the leaves to other parts of the plant. The phloem of flowering plants contains **sieve-tube members,** which have primary cell walls and living protoplasts at maturity. Associated with the sieve-tube members are the smaller **companion cells,** which play an important role in the transfer of substances from cell to cell. Phloem tissue may also contain sclerenchyma fibers and parenchyma.

Activity A: Vascular Tissue in Leaves

In the leaf cross section, xylem and phloem are found in vascular bundles that are scattered throughout the mesophyll. In a dicot leaf, veins run in all directions, so most of the vascular bundles have been cut at an angle. The best view of a cross section of a vascular bundle is the midrib, the major vein that runs through the center of the leaf. The most common arrangement of leaf vascular tissue is for the xylem to be located on the upper side of the vascular bundle and the phloem on the lower side. In this prepared slide, the xylem elements are stained red and have thick secondary cell walls, while the phloem cells are stained green due to their primary cell walls.

Procedure

1. Place the leaf cross section on your microscope.
2. Locate the vascular bundle at the midrib of the leaf.
3. On your previous drawing of the leaf cross section, sketch and label representative xylem and phloem tissue.

Activity B: Vascular Tissue in Roots

In the root cross section that you have been studying, the vessel elements are the thick-walled cells that form a cross in the center of the root. Since they have secondary cell walls, they are stained red.

The sieve-tube members and companion cells on the root cross section are stained green, since they possess primary cell walls. The phloem is located between the "arms" of xylem within the vascular cylinder delineated by the endodermis.

Procedure

1. Place the root cross section on your microscope.
2. Locate the central vascular cylinder. Identify the xylem and phloem.
3. On your previous drawing of the root cross section, sketch and label representative xylem and phloem cells in the vascular cylinder.

Activity C: Vascular Tissue in Stems

The vascular tissue in your *Coleus* stem cross section may be found in vascular bundles or it may be in a more or less continuous ring. The vessel elements are much larger than any other cells on the slide, so they are easily identifiable even if they are not differentiated from other cells by the stain. Notice again the thickness of the cell walls. The cells surrounding the vessel elements are also part of the xylem. The phloem is located just exterior to the xylem, toward the outside of the stem.

Procedure

1. Place the stem cross section on your microscope.

2. Locate a vascular bundle or, if the vascular tissue is continuous, an area of the slide that contains vascular tissue. Look for the large, thick-walled vessel elements. The phloem tissue is to the outside of the xylem. Unless you have a very good section, however, you may find it difficult to identify the phloem, which is easily crushed in the sectioning process.

3. On your drawing of the stem cross section from Exercise 17.2, sketch and label the xylem and phloem.

Notice that the vessel elements you have observed are hollow, since the protoplasts have died, and their cross-sectional area is very large compared to that of other cells. How are these structural features related to the function of the xylem?

In the stem, phloem is located exterior to the xylem. Explain why it is then logical that phloem in the leaf should be located below the xylem.

As the stem gains in girth, the vascular bundles are joined into a continuous ring, and the width of the band of xylem increases. The band of phloem, however, remains approximately the same width as old phloem cells are crushed by the new growth. Explain this difference between xylem and phloem. (Hint: Consider the cell walls.)

Questions for Review

1. If you are looking at a slide of a thin section of plant tissue, what clues might you have to tell you whether a cell has a primary or secondary cell wall?

2. Why do botanists use microtomes and stains when they want to look at plant tissues on the compound microscope?

3. What are the general functions of the following?
    Dermal tissue:

    Vascular tissue:

    Ground tissue:

4. How do the epidermal tissues of roots differ from those of stems and leaves? Suggest a reason for the differences.

5. Figure 17.4 shows the weight change in three apples over time. Which apple had its cuticle removed? Explain your answer.

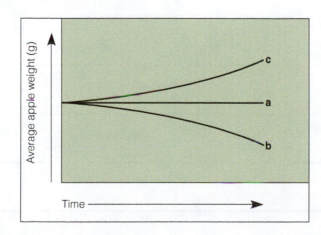

Figure 17.4.
Weight change in three apples.

6. How does the function of the parenchyma located in leaves differ from the function of the parenchyma located in the roots?

7. Describe how vessel elements are adapted for conducting water.

8. Summarize your knowledge of plant cells and tissues by filling in Table 17.1.

Table 17.1
Summary of Plant Cell and Tissue Types

	Living or Dead at Maturity	Primary or Secondary Wall	Location(s)	Function(s)
Epidermis				
Parenchyma				
Collenchyma				
Sclerenchyma fibers				
Sclereids				
Tracheids				
Vessel elements				
Sieve-tube members				
Companion cells				

Acknowledgment

The idea for using a nut and bolt for a microtome was suggested to me by Mrs. Rita Ouzts, a life science teacher at Irmo Middle School, Campus 1, Irmo, South Carolina.

Flowers, Fruits, and Seeds

Introduction

In seed plants, pollination is the mechanism by which sexual reproduction is carried out. The pollen produced by the "male" structures on one plant must somehow be transported to the "female" structures on another plant so that the sperm produced by the pollen grains can fertilize the eggs. The earliest seed-bearing plants, gymnosperms, were wind-pollinated, as are present-day pine trees. Pollen grains from male pine cones are carried by the wind to the female cones, where they land on a sticky resin. Insects feeding on other resinous parts of gymnosperms discovered this food source and began exploiting it; pollen stuck to their bodies as they fed. As they went from cone to cone and from plant to plant, pollination was also achieved. Since this method of pollen transfer was more effective than just wind, plants that were more attractive to insects had a better chance of being pollinated and reproducing. This is how the angiosperms evolved.

The flower, which is characteristic of the angiosperms (flowering plants), is an adaptation for effective pollination. The diversification of angiosperms and of insect pollinators such as bees and butterflies occurred, not surprisingly, during the same period of evolutionary history.

Figure 18.1 on the next page illustrates how the life cycle proceeds after fertilization. The fertilized egg and other tissues associated with it develop into a seed, while the ovary walls mature to become a fruit. The term angiosperm means "seed container." Angiosperms are not the only seed plants, but they are the only plants whose seeds are contained in an ovary. The fruit that develops may be dry, like peanut shells, or fleshy, like an orange. Besides offering protection, the fruit may have special modifications that help the seeds disperse to a suitable habitat.

In this lab topic, you will learn about different kinds of floral characteristics that attract different kinds of pollinators; you will examine various types of dispersal mechanisms used by plants; and you will look at the parts of a seed and the initial development of seeds into plants.

Outline

Exercise 18.1: Flower Structure

    Activity A: Flower Dissection

    Activity B: Pollination

    Activity C: Pollen Germination

Exercise 18.2: Fruits and Seeds

    Activity A: Classifying Simple Fleshy Fruits

    Activity B: Seed Dispersal

Figure 18.1.
Life cycle of a flowering plant.

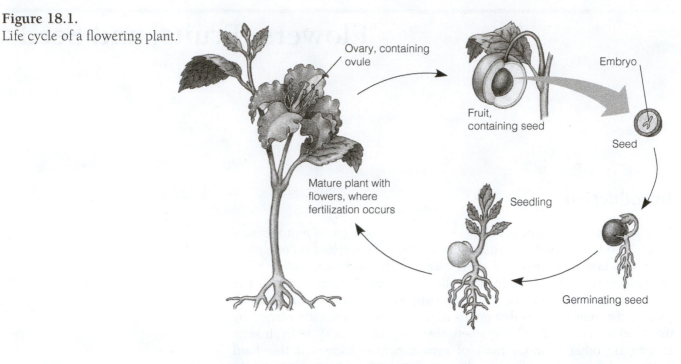

Ovary, containing
ovule

Embryo

Fruit,
containing seed

Seed

Mature plant with
flowers, where
fertilization occurs

Seedling

Germinating seed

Exercise 18.3: Seeds and Seedling Development
    Activity A: Seed Structure
    Activity B: Seedling Development

E X E R C I S E   1 8 . 1
Flower Structure

Objectives

After completing this exercise, you should be able to

1. Identify and describe the function of each part of a flower.
2. Identify flowers as complete or incomplete.
3. Distinguish between pollination and fertilization.
4. Discuss flowers as adaptations for effective pollination and name the floral characteristics preferred by the major pollinators.

Flowers are leaves that have been modified in various ways for reproduction. The fertile parts, which are directly involved in reproduction, are the stamens and the carpels. The sterile parts consist of the sepals and the petals. In general, the sepals protect the fertile parts and the petals attract pollinators. All of these parts are attached to the **receptacle**, a platform at the top of the flower stalk.

A flower that has all four parts (stamens, carpels, sepals and petals) is said to be **complete**. However, as you will see in Activity B, different species display diverse modifications of flower structure. Some flowers lack one or more of the four basic parts; they are called **incomplete**.

Activity A: Flower Dissection

The parts of the flower are arranged in concentric rings or rows around a structure called a receptacle at the top of the stem. From the outside to the inside, the rings are the sepals, petals, stamens, and carpels (Figure 18.2). In this activity you will dissect a flower and identify its parts.

Procedure

1. Get a flower from your instructor. Examine the sepals, petals, stamens, and carpels, as described in the following steps.

2. The outer ring of **sepals** protects the developing flower when it is a bud. When the flower opens, the sepals protect the fertile parts of the flower. The sepals are usually green, but in some flowers they have been modified to attract pollinators. In most members of the lily family, for example, the sepals are indistinguishable from petals except by their position in the outermost ring.

Some flowers also have showy leaves that are neither sepals nor petals. When you look closely at a dogwood "flower," you'll see that what appear to be large white petals are actually leaves surrounding a cluster of small yellow-petaled flowers. The poinsettia is another example; the red "petals" are actually leaves.

Remove the sepals after you have noted their position in the flower.

3. The next ring consists of the **petals,** which are generally colorful to attract pollinators. Petals will be discussed in more detail in Activity B.

Remove the petals after you have noted their position in the flower.

4. The male flower structures called **stamens** produce pollen grains, which are immature male gametophytes. As you may recall from Lab Topic 12 (Plant Diversity), the actual gametes (sperm nuclei) are not produced until after pollination (transfer of pollen to the female organ). The pollen is produced in the **anthers,** which are supported on a stalk.

Remove the stamens after you have noted their position in the flower.

5. The female flower structures are called **carpels,** and their function is to produce the female gametophytes (ovules), which then produce the female gametes (eggs). A flower may have one or more carpels, and if there is more than one they may be either fused or separate. At the top of the carpel is the **stigma,** which receives the pollen. The **ovary,** which encloses the ovules, is at the base of the carpel. If an ovule is fertilized, it matures into a seed. The ovary itself will become a fruit. Gymnosperms, which are also seed plants, have no ovary protecting the seeds, which is why they are called gymnosperms ("naked seeds").

6. Cut across the ovary with a razor blade as shown in Figure 18.2. The ovules are white and egg-shaped. How many carpels do you see?

7. Sketch the flower in the margin of your lab manual and label its parts.

Figure 18.2.
Structures found in a typical flower.

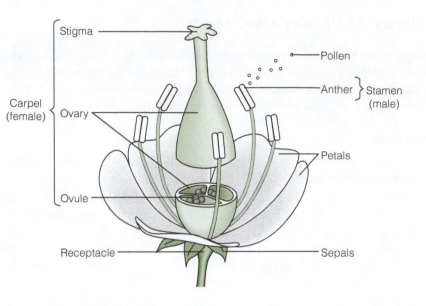

Activity B: Pollination

Animals visit flowers to obtain food, which may be pollen or nectar, a sugar-water. Most species of flowers have adaptations that induce pollination by particular animal species that, in turn, have adaptations tuned to the flowers they frequent. In this activity you will examine some general characteristics of the major pollinator and flower types.

Procedure

1. Read the following information about pollinators and complete Table 18.1 on page 18-6.

Beetles

Beetles were probably the first insect pollinators. From the plant's point of view, they are not the best pollinators, because they eat a lot of the pollen and chew on other plant parts as well. However, there are plant species that are solely beetle pollinated. The successful beetle flower has its ovules buried deep, out of harm's way. Because beetle vision is not very good, many beetle-pollinated flowers are white or dull-colored. Some flowers are large and single (magnolia, lily, California poppy, wild rose); others are small and clustered (dogwood, elderberry, spirea). Beetles can smell much better than they can see, so the flowers may have strong odors, usually fruity, spicy, or fetid rather than sweet. Some beetles feed on nectar, but it must be easily accessible because beetles lack the long sucking mouth parts of other pollinators. Color Plate 7 shows a beetle pollinator at work.

Bees

Nearly all of the 20,000 known species of bees visit flowers for food. They quickly learn to recognize colors, odors, and shapes so they can consistently use particular kinds of flowers. Bees can distinguish colors at the ultraviolet end of the spectrum that are invisible to us, but they cannot see red. Bee flowers are usually bright yellow, blue, or purple and they often have distinctive markings to lead bees to the nectar reward. Some of these "nectar guides" absorb light in the ultraviolet range and thus can only be revealed to the human eye by special techniques. Bee flowers usually supply nectar, which is made accessible only to insects with long sucking mouth parts such as bees have, and a landing platform for the bee is often provided by specially adapted petals, as shown in Color Plate 7. The flower odor is mild and sweet. Typical bee flowers are thistle, larkspur, snapdragon, violet, apple, heather, and the various mints.

Flies

There are two types of flies that visit flowers: long-tongued and short-tongued. The long-tongued flies are similar in adaptations and flower preferences to the bees, so we will only consider the short-tongued flies here. These insects have no particular specializations for feeding on flowers, and most have other sources of nourishment. They are attracted to flowers that smell like these other food sources, such as carrion, dung, humus, and blood. Flies do not have keen sight, so fly flowers tend to be dull colors, as illustrated in Color Plate 7. Often there is no nectar for them, and some flowers, such as Dutchman's-pipe, not only don't feed the fly, they trap it for a day or two while the fly receives a load of pollen. This pollen is transported to the next flower, where the fly will likewise be trapped so it can thoroughly pollinate the flower. Examples of fly-pollinated flowers are *Rafflesia* (said to smell like putrefying flesh), black arum (which smells like human dung), and a lily whose odor is compared to fish oil.

Butterflies

Many species of butterflies can see the entire color spectrum visible to us; butterfly flowers may be bright yellow, blue, red, or orange. Flowers are visited for their nectar, and the petals of some butterfly flowers form a long thin tube to store the nectar. Butterflies are successful in obtaining nectar from such a tube because their tongues are about 1–2 cm long (see Color Plate 7). Nectar guides are also common, and landing platforms are provided either by the single flower or by the closeness of a group of flowers. Butterflies are known to pollinate wild geranium, forget-me-not, pinks, and butterfly weed (an orange milkweed).

Moths

Moths often resemble butterflies and also have long tongues for taking up nectar. However, unlike butterflies and other pollinators (except bats) that are active during the day, most moths are nocturnal. Moths have a good sense of smell and typical moth flowers give off a strong, sweet odor, sometimes only at night. Many moth flowers remain closed during the bright daylight hours. In color, shades of white or other colors that stand out in the dusk are most common (see Color Plate 8). Tobacco, Easter lilies, and tuberoses are some moth- pollinated species.

Birds

In tropical and subtropical regions, pollination by sunbirds, hummingbirds, and honeysuckers is common (Color Plate 8). Birds have a poor sense of smell but good color vision, so bird flowers are odorless and red or yellow. Because birds have high metabolic requirements the nectar reward must be high. The abundant nectar is kept in tubes that are inaccessible to most smaller animals, and a red, odorless flower is effectively camouflaged from many insects. Examples of bird flowers are fuschia, hibiscus, cardinal flower, and bird of paradise.

Bats

Residents of temperate regions may be surprised to learn that many tropical species of flowers are bat-pollinated (Color Plate 8). Bat flowers are similar to bird flowers in being large and having abundant nectar. Bats, however, are nocturnal; the flowers they visit are white or dull-colored and may only open at night. Bat flowers have strong fruity or fermenting odors to attract bats. Some examples are calabash, sausage tree, candle tree, and organ-pipe cactus.

Wind

There are numerous wind-pollinated angiosperms that evolved from insect-pollinated ancestors. Wind flowers generally lack the accessories of attraction: They have no nectar, no odor, no bright colors, and are often without petals (Color Plate 8). The stamens are usually exposed to facilitate launching the masses of pollen, and the stigmas are likewise exposed to enhance chances of intercepting the pollen. Many wind-pollinated species have separate male and female flowers. Grasses, cottonwoods, oaks, and walnuts are among the groups of wind-pollinated flowering plants.

Table 18.1
Major Pollinators

Pollinator	Sensory Abilities	
	Visual	**Olfactory**
Beetles		
Bees		
Flies		
Butterflies		
Moths		
Birds		
Bats		

Table 18.2
Flower Modifications

Flower	Complete or Incomplete?	Beetles?	Bees?	Butterflies?	Birds?	Wind?	Other

2. Examine the flowers on display to see whether they are complete or incomplete, and record your observations in Table 18.2. Also comment on whether the pollinating agents listed in Table 18.2 could pollinate each flower, and explain why or why not.

Many species of flowers develop a special relationship with one type of pollinator, and pollinators may have reciprocal adaptations that enhance their "faithfulness" to a particular species of flower. Why would it be adaptive for plants to encourage exclusive pollinator service? (Hint: What happens if a pollinator visits a violet, then a daisy, and then an apple blossom?)

Activity C: Pollen Germination

As you could see when you dissected a flower in Activity A, the stigma is located at the other end of the carpel from the ovules. In some flowers this distance is as much as an inch or more. The microscopic pollen grain must germinate and produce a pollen tube that penetrates the stigma and grows through the entire length of the stalk that leads to the ovary. The pollen itself carries no food supply to provide the energy needed for this growth. Instead, cells on the surface of the stigma secrete a sugar-containing substance to provide nourishment. For many different types of plants, it is possible to germinate pollen in a drop of sucrose solution on a microscope slide.

Procedure

1. Using forceps, pluck a stamen from the flower provided. Hold the stamen by its stalk and brush the anther, face down, in the well of a concavity slide. (You may want to do this under a dissecting scope.)

2. Put a drop or two of sucrose solution in the well of the slide with the pollen. *Do not* use a coverslip.

3. Examine the slide using the scanning objective of the compound microscope. Sketch a few pollen grains in the margin of your lab manual.

4. Check your slide in 30 minutes to see if any germination has occurred. If the slide is drying out, add another drop of sucrose. Work on other exercises while you wait.

For the flower that you used for this activity, how far will the pollen tube have to grow in order to reach an ovule?

Why doesn't pollen carry its own food supply?

EXERCISE 18.2
Fruits and Seeds

Objectives

After completing this exercise, you should be able to

1. Define simple fruit, multiple fruit, aggregate fruit, pericarp, exocarp, mesocarp, and endocarp.

2. Using the key included in this exercise, classify a fruit as a berry, hesperidium, pepo, drupe, or pome.

3. Explain how fruits differ from vegetables.

4. Explain how fruits and seed coats are adapted to carry out seed dispersal.

5. Describe the characteristics of fruits and seeds that are dispersed by animals and by wind and give examples of each type.

Like flowers, fruits are unique to the angiosperms (flowering plants). The ovary first protects the female gametophyte and later the developing seeds. The ovary then develops into a fruit, which serves to protect and disperse the mature seeds. In some plants such as peaches the ovary wall is soft and fleshy when the fruit is mature. In others such as peanuts and grains the ovary wall is dry, papery, or hardened.

Activity A: Classifying Simple Fleshy Fruits

A **simple fruit** is a single ovary from a single flower. The ovary may, however, have more than one carpel, which is the chamber enclosing the seeds. For example, you can usually see three carpels in a cucumber slice (Figure 18.3).

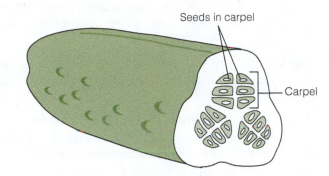

Seeds in carpel

Carpel

Figure 18.3.
Cucumber fruit showing three carpels.

A few fruits such as pineapple and fig develop from numerous flowers that grow very close together; their ovary walls fuse during maturation. Fruits that develop from multiple flowers are called **multiple fruits.** An **aggregate fruit** is a group of simple fruits that are joined together. Aggregate fruits develop when a single flower has numerous ovaries attached to the same flower. Each ovary develops into a separate fruit, but since the fruits are all attached to a single receptacle (the platform at the tip of the stem) they appear to be a single fruit. Blackberries and raspberries are examples of aggregate fruits: each "knob" of a blackberry contains a seed. Strawberries are also aggregate fruits: each "seed" is actually a dry ovary (simple fruit) containing a seed.

As the ovary of fleshy fruits ripens and grows, it also differentiates into three layers. Collectively, the layers are called the **pericarp.** "Carp" means "fruit," so the outer layer of the fruit is called the **exocarp,** the inner layer is the **endocarp,** and the middle layer is the **mesocarp.** Figure 18.4 illustrates the pericarp of a peach. The fuzzy skin is the exocarp, the juicy middle is the mesocarp, and the stony pit is the endocarp. The seed is inside the endocarp, which protects the seed as it passes through an animal's digestive tract. The differences among the pericarps of various fleshy fruits are useful for classifying them.

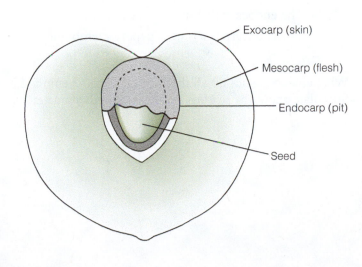

Exocarp (skin)

Mesocarp (flesh)

Endocarp (pit)

Seed

Figure 18.4.
The layers (pericarp) of a peach fruit.

When biologists want to identify an organism, they often turn to a classification scheme known as a **key.** There are keys to identify trees, mushrooms, seashells, mammals, insects, and most other groups of organisms. Keys that are used by experts may include thousands of species, but less complex keys have also been constructed for use by nonscientists who simply want to know more about nature. A key is constructed to lead you through a series of comparisons. At each step, you choose the statement that best describes the sample you are trying to identify. As this process is repeated, the classification is increasingly narrowed until you obtain an answer.

The key that you will use in this exercise includes only fleshy fruits. Dry fruits such as peanuts, beans, grains, and nuts cannot be identified using this key.

To use the key, examine each fruit and choose between the two alternatives for number 1: Is the fruit better described by 1a or by 1b? Continue with the choices listed below the alternative that you selected. For example, if you selected 1a, your next choice is between 2a and 2b below 1a. A peach would key out as follows: The endocarp is not fleshy, since it is the stony pit: choose 1b. The endocarp is stony rather than papery: under 1b, choose 4a. This identifies the fruit as a drupe.

As another example, think about a pumpkin. The endocarp is fleshy (when you scoop out the inside of a pumpkin, the seeds are not separated from the rest of the fruit by stony or papery tissue): choose 1a. The exocarp is thickened: choose 2b. The exocarp is not easily peeled away from the rest of the fruit (have you ever tried to peel a pumpkin?): choose 3b. The pumpkin is a pepo.

Key to Simple Fleshy Fruits

1a. The endocarp is fleshy or pulpy. . . . . Go to 2

    2a. The exocarp is thin and skinlike. . . . . **Berry**

    2b. The exocarp is thickened. . . . . Go to 3

        3a.The exocarp and mesocarp are leathery and easily peeled away from the rest of the fruit. The carpels contain large juice-filled hairs. . . . . **Hesperidium**

        3b.The exocarp is thick but not leathery and cannot easily be separated from the rest of the fruit. . . . . **Pepo**

1b. The endocarp is not fleshy. . . . . Go to 4

    4a. The endocarp is stony. . . . . **Drupe**

    4b. The endocarp is a tough, papery membrane. . . . . **Pome**

What characteristics do all the fruits that you examined have in common? Speculate about the reasons for these shared characteristics.

What are the major differences among the fruits that you examined?

What Is a Vegetable?

Some of the fruits you keyed out are better known to you as vegetables. There is no botanical definition of a vegetable.

What is your definition of a vegetable?

List some other vegetables.

Which of the vegetables you listed could be classified as fruits?

Based on the botanical definition of fruit, what's the best way to distinguish a fruit from a vegetable?

Activity B: Seed Dispersal

Many fruits and seed coats (the layer of tissue covering the seed) are adapted for ensuring that the seeds are carried away from the parent plant, a process known as seed dispersal.

The modifications for seed dispersal parallel the modifications of flowers for pollination. Animal agents may be employed, and plants use color and nutritional rewards to attract the animals to their fruits. Fleshy fruits such as peaches, for example, are meant to be eaten. The animal digests the fruit's flesh, but the seed, in this case protected within the endocarp, passes unharmed through the gut and is deposited some distance from the parent plant. In other fruits, animals consume some seeds while others are buried and forgotten. Name some fruits that use animals to disperse their seeds.

Animals are more passive participants in another scheme of seed dispersal, in which the seed coats are adapted to latch onto the fur of passing animals. These plants have names that describe the seed coats, such as sticktights and beggar-ticks.

Other plants depend on dispersal by the wind. Typically they produce many light seeds, and the seed coat forms some sort of structure that can easily be picked up and carried along by the breeze. Name some examples of seeds that are dispersed by wind.

A few seeds are also adapted for dispersal by water, and the fruits of others are simply dropped and carried away passively by rainwater or wind without any special mechanisms.

Procedure

1. Examine the fruits and seeds on display and determine the probable means of dispersal.
2. In Table 18.3, record the name and probable dispersal method of each plant, and list the characteristics that helped you decide on the dispersal method.

Table 18.3
Seed Dispersal

Plant	Dispersal Method	Characteristics

Why is it advantageous for plants to disperse their offspring?

In Activity A you listed the similarities and differences that you observed among the fleshy fruits you examined. Suggest reasons for those characteristics.

E X E R C I S E   1 8 . 3
Seeds and Seedling Development

Objectives

After completing this exercise, you should be able to

1. Identify the parts of a seed and describe the function of each part.
2. Identify the parts of a seedling.
3. Compare the development of corn and bean.

Activity A: Seed Structure

The ovules inside the ovary are the female gametophytes. Each ovule produces an egg. If the egg is fertilized by a sperm nucleus that has been produced by the pollen (male gametophyte), it develops into an embryo. Recall from Lab Topic 12 (Plant Diversity) that flowering plants have double fertilization. A second sperm nucleus fertilizes two additional nuclei to form the **endosperm,** which becomes a food source for the embryo when it germinates. The ovule is also surrounded by a layer of tissue that becomes a **seed coat** to protect the seed. In this activity you will look at the three parts of a seed.

Procedure

1. Get a peanut fruit from your instructor.
2. Crack and remove the fruit (shell) from the seeds. As you do, note that the peanut shell can easily be split in half lengthwise. It is a pod, similar to the fruits of other members of the bean family.
3. Remove the red, papery covering, which is the seed coat. Layers of tissue surrounding the ovule thicken and harden into this protective covering.
4. Open the **cotyledons** (seed leaves) gently to avoid damaging the rest of the embryo. The halves of the peanut are the cotyledons. The peanut cotyledons have absorbed the endosperm and will now serve as the food supply. In some plants, for example corn, the endosperm remains separate from the cotyledon(s).

5. Observe the **embryo.** The cotyledons are actually part of the embryo. The rest of the embryo is the root/shoot axis; the plant will grow from the root tip and the shoot tip. Notice the first "foliage" leaves apparent in the embryo.

6. In the margin of your lab manual, sketch and label the parts of the seed.

Activity B: Seedling Development

The molecular structure of the seed coat makes it especially effective at taking up water. As water is absorbed, enzymes are activated that convert the stored starches or fats in the endosperm or cotyledons into usable food molecules. The initial growth of the embryo, which is called germination, depends on the metabolic energy derived from this food. Later, photosynthesis will provide energy for the plant to continue its growth.

Flowering plants are divided into two subgroups, monocotyledons and dicotyledons, depending on whether there is one cotyledon or two in the seed. In this activity you will compare the seeds and early development of a monocot and a dicot.

Procedure

1. Get a bean seed, a corn seed, and a plastic cup containing bean and corn seedlings from your instructor.

2. Dissect the bean seed by removing the seed coat and separating the cotyledons to reveal the embryo inside. Beans are in the same plant family as peanuts.

Are beans monocots or dicots? How do you know?

3. In the margin of your lab manual, sketch the seed and label the seed coat, cotyledons, and embryo.

4. Make a sketch of the bean seedling in the margin of your lab manual.

> ❋ **Since the number of seedling sets is limited, please do not dig up and destroy these plants!**

5. Label the cotyledons on your sketch of the bean seedling.

6. Also label the **hypocotyl** (the part of the shoot below the cotyledons) and the **epicotyl** (the part of the shoot above the cotyledons) on your drawing of the bean seedling. The two leaves at the top of the epicotyl are known as the **first foliage leaves,** in contrast to the cotyledons, or seed leaves.

7. Label the **primary root,** which is the largest root. The primary root will become a taproot, growing directly downward through the soil. You can see lateral roots growing out of the taproot; they will branch out in all directions to utilize a greater area of the soil beneath the plant, thus increasing the surface area for absorption of water and minerals.

The bean cotyledons may be somewhat shriveled when you observe the seedling. Why are the cotyledons no longer as fleshy as they were in the seed?

If you observe seed germination, you will notice that the root emerges from the seed coat first, followed by the shoot. Why should the root develop more quickly? (Hint: What are the functions of the root?)

8. Cut open a corn seed longitudinally with a razor blade.

⚠ Be careful not to let the razor blade slip. Even though the corn has been soaked in water to soften it, it is very hard to cut.

Corn, like other grains, has a dry fruit that is fused with the seed coat inside, so it is not possible to distinguish the ovary wall from the seed coat.

9. Compare the corn seed you have dissected to Figure 18.5.

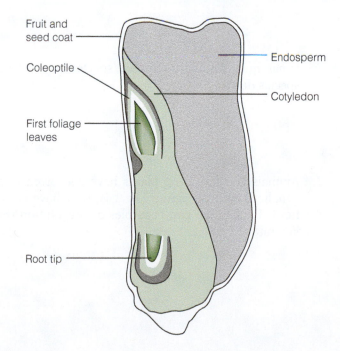

Fruit and seed coat
Coleoptile
First foliage leaves
Root tip
Endosperm
Cotyledon

Figure 18.5.
Corn seed. The endosperm in corn is not absorbed by the cotyledon. Notice the coleoptile, which is a sheath enclosing the leaves.

10. Examine the corn seedling and sketch it in the margin of your lab manual.

11. Label the **first foliage leaf;** it should be apparent. The emerging leaves are encased in a sheath called the **coleoptile.** The coleoptile is found in all members of the grass family (which includes corn and the other grains), but it is not a characteristic of monocots in general.

12. Examine the roots and label them on your drawing.

The primary root lasts for only a short time in monocots; the entire root system develops from the end of the stem. (This is similar to how roots arise from stem cuttings placed in water or moist soil.) If you have ever grown corn in a garden, you have probably noticed how the roots grow from the stem. Mounding soil high up on the stalks stimulates root growth, thus improving the plant's support. Since there is no taproot to penetrate deep into the ground, the roots of monocots tend to be rather shallow and spread out.

Which type of plant do you think is easier to transplant, a dicot or a monocot? Explain.

Questions for Review

1. Some angiosperm species have returned to the wind-pollinated mode of gymnosperms (for example, pines). There are benefits and costs associated with both strategies. It costs (energy and materials) to make pollen. It also costs to use animal go-betweens. What must wind-pollinated species "pay" for that animal-pollinated species don't?

What must animal-pollinated species pay for that wind-pollinated species don't?

2. Animals and flowering plants have coevolved in some ways that are beneficial to both parties. That is, they have complementary adaptations. Explain the two examples of coevolution that were seen in this laboratory.

3. An imaginary animal has a poor sense of smell, a long tongue (1–2 cm), and sees well at the short-wavelength end of the spectrum (ultraviolet and blue) but not at the other end of the spectrum (red). Design a flower that attracts this pollinator *and* excludes other pollinators.

4. Many fruits are green when they are immature and turn red or orange when they are ripe. What is a possible reason for this?

5. Keys like the one you used to classify fruits are based on characteristics that distinguish one group from another. In this case, only the layers of the pericarp were used. What other characteristic of simple fleshy fruits would be useful in a key?

6. How do you know that zucchini is a fruit as well as a vegetable?

7. Occasionally in nature a mutant plant that lacks the ability to produce chlorophyll occurs. How long could such an "albino" plant survive, if at all? Explain.

8. How might the amount of food supply that a seed contains be related to its dispersal mechanism?

9. What seed characteristic should be considered in deciding how deep to plant a seed? Explain.

Animal Behavior

Introduction

In Lab Topic 16 (The Sensory System), you learned about how animals perceive the environment. Sensory input is processed by the central nervous system, which then directs a response. **Behavior** is the term given to those responses. Behavior may be simple and reflexive, such as rapid withdrawal from a source of pain, or it may be complex, such as detecting, approaching, and courting a potential mate. In short, behavior includes everything animals do.

Human behavior is governed by emotion as well as intellect and reasoning skills. Since we are capable of understanding our own behavior, we often find it difficult to adopt a scientific perspective on animal behavior. For many of us, association with pets has created a most unscientific tendency to anthropomorphize, or attribute human thoughts and feelings to animals. For example, when a kitten pounces on a ball of yarn and tumbles it around, we might characterize its behavior as recreation, as if the kitten intentionally decides to relax and have a bit of fun. In fact, this play behavior is probably a rehearsal for the cat's predatory habits later in life. The ball of yarn is a stand-in for a field mouse, and the kitten is learning to chase, capture, and subdue its prey.

As in any scientific field of study, we must be able to formulate testable hypotheses in order to answer questions about behavior. Hypotheses about the causes of behavior can address two general types of questions: What triggers the behavior? and What is the adaptive significance of the behavior? For example, the cue, or trigger, for the kitten's attack on the ball of yarn is whatever stimulus or stimuli caused the ball to be perceived as prey. It might be the ball's shape, color, smell, or movement. The evolutionary reason, or adaptive significance, for the behavior explains why an animal engages in a particular behavior. For example, juvenile rehearsal of prey capture may have provided an evolutionary advantage for the wild ancestors of our domestic cats.

In this lab topic you will investigate a relatively simple behavior, kinesis in pill bugs, and a more complex behavior, agonistic encounters between bettas (Siamese fighting fish). The methods employed allow you to study the immediate causes of these behaviors, but you should also consider their adaptive significance.

Outline

E X E R C I S E   1 9 . 1
Kinesis in Pill Bugs

Objectives

After completing this exercise, you should be able to

1. Define kinesis.
2. Describe how a choice chamber is used for determining pill bug environmental preferences.
3. Explain how the behavior of pill bugs contributes to their survival.

Most humans have sense enough to come in out of the rain, but even lower animals engage in behaviors that will place them or keep them in favorable environments. In one such behavior, called **taxis,** animals move directly toward or away from a stimulus. In another type of behavior, called **kinesis,** animals move randomly in response to environmental stimuli. Their movements are relatively rapid when the environment is unfavorable and slow down when the environment is favorable. How could this random movement be useful to an organism in finding a suitable place?

What environmental cues might cause kinetic behavior?

What might be the adaptive significance of kinetic behavior?

Figure 19.1.
The pill bug. Other common names for the pill bug and related species are roly-poly and sow bug.

The pill bug (Figure 19.1) is an arthropod belonging to the class Crustacea, which also contains lobsters, crabs, and barnacles. It is one of the few land-dwelling members of this class. You can usually find pill bugs by turning over logs and rocks in the woods or even in your backyard.

The kinetic response of pill bugs to different environmental stimuli can be investigated using an apparatus called a choice chamber. As shown in Figure 19.2, the choice chamber consists of two large petri dishes taped together with an opening in between. Each chamber can have different conditions. To determine the response of pill bugs to environmental conditions, place an equal number of pill bugs in each dish. Observe their movements and record their "choice" of chamber conditions.

Covers (upside down)

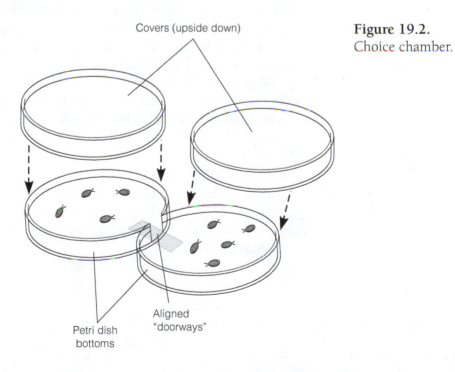

Petri dish
bottoms

Aligned
"doorways"

Figure 19.2.
Choice chamber.

Hints for Successful Investigations

1. Use a soft brush when you transfer pill bugs from one container to another.

2. Once you have started an experiment, don't disturb the chamber.

3. When you get pill bugs from the stock culture, avoid picking the very small ones. They tend to ride on the backs of the larger ones instead of moving about themselves.

4. If your pill bugs all pile up in one chamber and stop at the beginning of an experiment, they are probably tired from previous experiments and are not responding to environmental stimuli. Put them in the "recovery" terrarium and replace them with a fresh batch from the stock culture.

In the first activity, conditions in both chambers will be exactly the same. This exercise will make you familiar with the general behavior of the pill bugs and with the technique for recording data on their behavior.

Predict the distribution of the pill bugs after 5 minutes.

Procedure

1. Put a piece of filter paper in a regular-sized petri dish and moisten it with a wash bottle. You will use this dish to transport pill bugs.

2. Use a soft brush to transfer 10 pill bugs from the stock culture into your dish.

3. Set your choice chambers on the bench in front of you. Environmental conditions should be the same in both chambers, so be sure they rest on a similar color background. Be careful not to let your shadow fall over either side once the pill bugs are in the chambers.

4. Use a soft brush to transfer five pill bugs from your petri dish into each chamber. Quickly put the lids over the chambers. (Put the lids on upside-down or they won't fit.) Note the position of the second hand on your watch or the wall clock.

5. At 30-second intervals, count the number of pill bugs in each chamber and note their activity level, using phrases such as "very active," "some activity," or "hardly moving" to describe what the pillbugs are doing. Record this information in Table 19.1.

6. Continue your observations for 5 minutes.

7. Return the pill bugs to your petri dish.

8. Use Figure 19.3 to graph the number of pill bugs present in Chamber 1 during the 5-minute observation. Label the axes and write a descriptive caption for the figure.

Table 19.1
Observations of Pill Bug Location and Activity Level
(five pill bugs placed in each chamber initially; chambers
have identical environmental conditions)

Minutes	Chamber 1		Chamber 2	
	Number of Pill Bugs	Activity	Number of Pill Bugs	Activity
0				
0.5				
1.0				
1.5				
2.0				
2.5				
3.0				
3.5				
4.0				
4.5				
5.0				

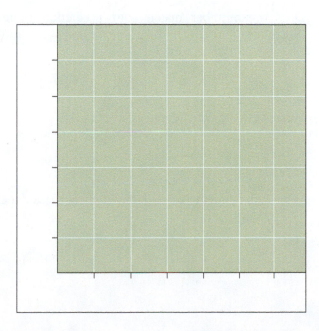

Figure 19.3.

Describe and attempt to explain the general pattern of movement when pill bugs are active.

How do pill bugs behave when they are inactive?

After 5 minutes, how were the pill bugs distributed in the two chambers? Did this distribution match your prediction?

9. The instructor will collect the results from all the groups in the class. Obtain the average number of pill bugs in Chambers 1 and 2 after 5 minutes and record the results in Table 19.2.

Table 19.2
Number of Pill Bugs in Each Chamber
After 5 Minutes (class average)

Chamber 1	Chamber 2

Does this number match your prediction more closely? Explain the difference.

What were the dependent variables in this experiment?

What independent variables might affect these dependent variables?

EXERCISE 19.2
Agonistic Behavior in Bettas

Objectives

After completing this exercise, you should be able to

1. Define agonistic behavior and explain its adaptive significance.
2. Give examples of animal behaviors that indicate aggression and submission.
3. Explain what a fixed action pattern is and how it is triggered.
4. Describe the agonistic behavior of bettas.

Members of the same species living in the same area are in competition for the same resources, including food, water, space, shelter, and mates. Many animals have developed **agonistic** behaviors for encounters with these rivals. ("Agonistic" is derived from a Greek word meaning struggle or contest.) Essentially, the competitors stage a contest to determine which one gets the prize. Animals may stake out territories that they then defend against intruders. Many birds do this, and often when you hear a bird singing you are listening to it advertising its territorial claim.

In most species, these contests are highly ritualized and symbolic, involving more posturing than fighting. Even in species that will engage in combat, however, the contestants first attempt to win the fight through intimidation by presenting themselves as large, fierce, and unbeatable. One participant may then perceive itself to be inferior and back down.

Give an example of an aggressive or threatening display.

The loser in an agonistic encounter may display submissive or appeasement behavior and thus acknowledge the other animal's superiority. This generally halts the aggressive display on the winner's part.

Give an example of submissive behavior.

What is a possible adaptive significance of agonistic behavior?

In this exercise you will investigate agonistic behavior in Siamese fighting fish (Figure 19.4). The scientific name for these fish is *Betta splendens*, so they are often referred to as bettas (pronounced bay-tuh). These fish, especially the males, are territorial. The presence of an intruder triggers a display that is intended to intimidate the rival.

Figure 19.4.
Siamese fighting fish (*Betta splendens*).

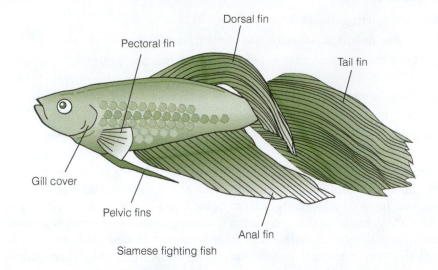

The agonistic behavior of the betta is inborn and follows a fixed action pattern (FAP). That is, it goes through a predictable sequence once the behavior is triggered by an innate **releasing mechanism**, or cue. Another example of a FAP is when a parent bird returns to the nest bearing food. The young birds open their mouths wide and chirp noisily. That behavior triggers the parents' feeding behavior, and the parents stuff the food down the throats of their young. A mechanical analogy for a fixed action pattern is a coffee vending machine. When you put money in (the releasing mechanism), the machine drops a cup into place and then pours coffee into it. If the cup dispenser is empty, the coffee is still poured: Once the mechanism is triggered, it goes through its entire sequence of behaviors.

What is a possible adaptive significance of having a releasing mechanism?

Activity A: Solitary Behavior of the Betta

In this activity you will observe a betta at rest. Your purpose is to learn to identify the physical and behavioral characteristics of the fish when it is undisturbed so you will have a basis for comparison when the betta encounters a rival.

Procedure

1. Set the fish tank in a place where all the group members can comfortably observe the fish and where it is not near another betta. Bright-colored clothing may trigger a display, so keep your distance.

2. Observe the betta's appearance and behavior and note them below. In particular, describe the fish's color, the position of its fins, and its movement around the tank. Do not disturb the fish in any way while you are observing it.

Activity B: The Betta's Reaction to Its Mirror Image

In this procedure you will show your fish its own reflection. Predict how the fish will react.

Procedure

1. Hold the mirror against the side of the tank.

2. Observe how the fish responds and record your observations in the space below. Again, pay particular attention to the color, fins, and movements.

 It may take a few minutes before the fish notices the "intruder," so be patient.

How does the display behavior affect the appearance of the fish?

Name another animal that uses a similar effect when confronted by an enemy.

What are some possible releasing mechanisms for the betta's display behavior?

E X E R C I S E 1 9 . 3
Designing an Experiment

Objective

After completing this exercise, you should be able to

1. Design an original experiment to investigate some factor that affects kinesis of pill bugs or agonistic behavior of bettas.

In Exercises 19.1 and 19.2 you learned techniques for investigating the kinetic behavior of pill bugs and agonistic behavior of bettas. In Exercises 19.3 and 19.4 your lab team will design an experiment using one of these methods, perform your experiment, and present and interpret your results. You may want to review the factors that you listed as possible causes for these behaviors to help you decide on an independent variable for your investigation.

Your instructor will tell you which animal to use for your investigation. If you are doing a pill bug experiment, you will use the choice chamber that you used in Exercise 19.1. If you are doing a betta experiment, you will use the fish that were used in Exercise 19.2. Your instructor will tell you what additional materials will be available.

Describe your experiment below.

Question or Hypothesis

Dependent Variable

Independent Variable

Explain why you think this independent variable will affect the dependent variable.

Control Treatment(s)

Replication

Brief Explanation of Experiment

Predictions

What results would support your hypothesis? What results would prove your hypothesis to be false?

Method

Include the levels of treatment you plan to use.

Design a Table to Collect Your Data

List Any Additional Materials You Will Require

E X E R C I S E   1 9 . 4

Performing the Experiment and Interpreting the Results

Objectives

After completing this exercise, you should be able to

1. Perform the experiment your lab team designed.
2. Present and interpret the results of your experiment.

Before you do the experiment, be sure that everyone on your lab team understands how the experiment will be performed.

Be thorough in collecting data. Don't just write down numbers; record what they mean as well. Don't rely on your memory for information that you will need when reporting on your experiment later! If you have any questions, doubts, or problems during the experiment, be sure to write them down, too.

Results

Before you begin to prepare your results for presentation, decide on the best format to use. Remember, you want to give the reader a clear, concise picture of what your experiment showed. Refer to the data presentation section of Appendix A (Tools for Scientific Inquiry) for help. If you are drawing graphs, use graph paper. Complete your tables and/or graphs before attempting to interpret your results.

Write a few sentences below *describing* the results (don't explain why you got these results or draw conclusions yet).

Discussion

Look back at the hypothesis or question you posed in this experiment. Look at the graphs or tables of your data. Do your results support your hypothesis or prove it false? Explain your answer, using your data for support.

Did your results correspond to the prediction you made? If not, explain how your results are different from your expectations and why this might have occurred.

Describe how your data are supported by information from other sources (for example, textbooks or other lab teams working on a similar problem).

If you had any problems with the procedure or questionable results, explain how they might have influenced your conclusion.

If you had an opportunity to repeat and extend this experiment to make your results more convincing, what would you do?

Summarize the conclusion you have drawn from your results.

Questions for Review

1. How do pill bugs locate appropriate environments?

2. If you suddenly turn over a rock and find pill bugs there, what do you expect them to be doing? If you watch the pill bugs for a few minutes, would you expect to see their behavior change?

3. Describe an experiment to test whether pill bugs are attracted to tannic acid (a chemical found in rotting wood).

4. How does agonistic behavior contribute to an animal's survival and reproduction?

5. Most agonistic behavior involves ritualized display rather than actual combat. Give a political analogy for this behavior.

Foraging Strategy

Introduction

One essential animal behavior is foraging for food, and many structural and behavioral adaptations of animals are linked to efficient food gathering. If an animal eats other animals, it must hunt, recognize, capture, and subdue the prey, activities that clearly require complex behaviors. An animal that uses plants as food resources must be able to locate appropriate plants. This may seem to be a simple task since plants are not motile, but it isn't necessarily as easy as it sounds. Consider a bee that feeds on nectar or pollen, for example. Flowers of the right species may be located at some distance from each other, so the bee must expend energy to locate and travel between these food resources, and it must be able to recognize the flower when it finds it.

Ecologists often use models to help them understand complex interactions such as those involved in foraging behavior. The purpose of a model is to allow the scientist to simplify and define the components of a real process, and thus be able to study each component separately. It is a tool for studying the natural world in a controllable setting. You will use a type of model called a simulation to investigate aspects of foraging behavior.

Outline

EXERCISE 20.1
The Simulation System

Objective

After completing this exercise, you should be able to

1. Describe the components of the simulation used in this lab topic and explain how they represent real processes.

In this simulation the food resources (prey) are represented by dried beans. Variety in the size and color of beans allows us to simulate having different types of food items. The food items are located in a pan of sand, which represents a patch of habitat.

We must also choose an animal to do the foraging. Ideally, the animal should be able to follow directions so that the rules of the simulation are observed. The animal should be intelligent enough to apply several variations on the basic pattern. We have therefore decided to use *you* in the role of the foraging animal.

What physical and mental abilities do humans have as foragers or predators?

What animals share some or all of these characteristics?

Like scientific investigations, simulations are defined by their rules. If each participant in the simulation follows the rules, we will be able to draw valid conclusions from the data.

The Rules

1. You may use only one hand to forage.

2. You may not remove *any* sand from the pan as you forage.

3. Each member of the group forages in turn. A forager has a maximum of 10 seconds to find enough prey to survive.

4. Each prey item must be placed in a dish next to the pan before the forager may capture the next prey. This simulates handling time, which is the time that the forager requires to subdue and consume its prey.

5. One team member should act as the timer/recorder/referee while another forages. After each turn, the recorder should smooth over the sand and cover any prey that have been revealed before the next person forages. The recorder also serves as the referee and should verify that the forager follows the rules. Rotate this job so that each person who forages becomes the recorder for the next group member.

6. Only the recorder may watch the person who is foraging.

7. When you have finished collecting data from a simulation, pour the sand through the sieve into the empty pan and remove any beans that remain.

EXERCISE 20.2
Single Prey Species

Objectives

After completing this exercise, you should be able to

1. Describe a successful foraging strategy for this simulation.

2. Give examples of natural situations to which the simulation in this exercise might apply.

To demonstrate the simulation, only one prey species, kidney beans, will be available to the foragers. You should work in teams of four or five.

Procedure

1. Disperse 30 kidney beans in the sand. All beans should be covered so they are not showing on the surface.

2. Team members will take turns foraging. You may not watch each other forage (except for the timer/recorder/referee). Each forager must find three kidney beans in a maximum of 10 seconds in order to survive. The forager should stop after finding three beans. A forager who doesn't find three beans within 10 seconds dies of starvation.

3. Record in Table 20.1 the number of prey remaining after each forager's turn and the time it actually took for each forager to find the three beans.

4. Continue taking turns foraging until each team member has died.

5. Record in Table 20.1 the total number of foraging turns taken in the simulation.

What data measure the success of the foragers?

What happens as a food resource is depleted?

What real situation might this simulation apply to?

Table 20.1
Results of Foraging Simulation. Each forager has 10 seconds to capture three prey. The only prey species is kidney beans.

Forager	Number of Prey Remaining	Time to Find Three
1	30	
2		

Total number of foraging turns in this simulation: _____

What do animals do when food resources are depleted in nature?

How many beans were left in the sand when all the foragers died? How does this result relate to a natural situation?

Besides observing other animals directly, how might animals avoid wasting time in an area that has already been searched by another animal?

From the forager's point of view, is the kidney bean an easy prey item to capture? Explain.

Observe the other prey items that are available and speculate on how the differences between them and kidney beans might make them easier or more difficult to capture.

Great Northern beans:

Lima beans:

Lentils:

How might you change the simulation to improve the forager's success in capturing prey?

How might you change the simulation to increase prey survival?

EXERCISE 20.3
Mixed Prey Species

Objectives

After completing this exercise, you should be able to

1. Distinguish between specialist and generalist foragers and give examples of each type.

2. Give examples of natural situations to which the simulation in this exercise might apply.

In Exercise 20.2, only one type of prey was available. The forager had to find this prey or it would die. This type of forager is called a **specialist.** Many herbivorous animals specialize in eating certain types of plants. Usually specialists will eat a number of related species, but a few have extremely restricted diets. Name an example of a specialist. (Hint: Think of warm, fuzzy herbivores.)

Generalists, on the other hand, are able to exploit a variety of food resources. The raccoon, for example, is an opportunistic generalist. Besides its natural diet of fish, frogs, and crayfish, a raccoon eagerly consumes hot dogs, potato chips, cole slaw, and whatever else is not securely locked up by campers.

What is an advantage of being a specialist?

What is a disadvantage of being a specialist?

What is an advantage of being a generalist?

What is a disadvantage of being a generalist?

All food is not of equal value. Some prey items are larger than others, and thus the forager obtains more food per capture. The food resources may differ in nutritional value. For example, one seed may contain carbohydrates while another seed of the same size contains fat, which has a higher caloric value per gram than carbohydrate. However, the forager's choice of food is also influenced by the abundance of the resource. Food A may contain more calories than Food B, but it may be so rare that it isn't worth the extra energy the animal must expend to find it.

Predators develop a "search image" of the prey they are seeking. We do something analogous when we are looking for something. It helps to have some feature to key in on. For example, when you're looking for a person in a crowd, it's useful to know what the person is wearing. Once a forager has a search image for kidney beans, it is easier to find that prey than switch to another.

In this simulation there are three different types of prey available: kidney beans, Great Northern beans, and lima beans. The population size is different for each species. The number of prey each forager must find in order to survive is also different. There are 40 kidney beans, 30 Great Northern beans, and 20 lima beans.

The food values of the various items are different. A forager survives if he or she can find any of the following combinations:

4 kidney beans	3 kidney beans and 1 Great Northern bean
3 Great Northern beans	
2 lima beans	3 kidney beans and 1 lima bean

Are the foragers in this simulation specialists or generalists?

Procedure

1. Sieve the sand to remove any prey remaining from the previous simulation.

2. Disperse 40 kidney beans, 30 Great Northern beans, and 20 lima beans in the sand.

3. Take turns foraging, observing the same rules that were used in the previous simulation. Each forager must find one of the specified combinations of prey within 10 seconds in order to survive. The forager should stop after finding enough prey to survive. A forager who can't capture enough prey starves to death. After each person forages, he or she is the recorder/referee for the next forager.

4. In Table 20.2 record the number of prey remaining after each forager's turn, the number of seconds it takes each forager to capture the right combination of beans, and the total number of foraging turns before all the foragers die.

Table 20.2
Results of Foraging Simulation. Each forager has 10 seconds to capture enough prey for survival. The three prey species—kidney beans, lima beans, and Great Northern beans—have different food values.

Forager	Number of Prey Remaining			Number of Prey Captured			Time (sec)
	Kidney	Great Northern	Lima	Kidney	Great Northern	Lima	
1	40	30	20				
2							

Total number of foraging turns in this simulation: _____

What strategy did you use in deciding which prey to look for?

Would your strategy have been different if you hadn't known the initial number of each prey present? Explain.

Was one of the resources depleted first? If so, which one?

What happened after this prey species was depleted?

Suppose each forager had specialized on kidney beans, lima beans, or Great Northern beans. How might the results have been different?

EXERCISE 20.4
Designing an Experiment

Objective

After completing this exercise, you should be able to

1. Design a variation on the simulation used in Exercises 20.2 and 20.3 to investigate some aspect of foraging strategy.

Exercises 20.1, 20.2, and 20.3 explained and demonstrated a method of simulating foraging strategy by using beans as the prey and students as the foragers. In Exercise 20.4 your lab team will design a variation on this simulation to investigate some other aspect of foraging strategy. In Exercise 20.5 your lab team will perform the simulation and present and interpret your results. You may want to review the questions you answered for Exercises 20.2 and 20.3 to decide on an independent variable for your investigation. Possible dependent variables include the length of time required by each forager to find enough prey to survive each turn, the total number of foraging turns in a simulation, the number of prey remaining after each round of foraging, or a combination of data that are useful for evaluating your hypothesis. Use the format below to plan and describe your simulation.

For your investigation, you will use the pan of sand that you used for Exercises 20.2 and 20.3. Kidney beans, lima beans, Great Northern beans, and lentils are also supplied. Your instructor will tell you if any additional materials are available.

Hypothesis

Dependent Variable(s)

Independent Variable

Explain why you think this independent variable will affect the dependent variable(s).

Replication

Predictions

What results will support your hypothesis? What results will prove your hypothesis to be false?

Method

Explain how you have modified the rules of the simulation.

Design a Table or Tables to Collect Your Data

EXERCISE 20.5
Performing the Experiment and Interpreting the Results

Objectives

After completing this exercise, you should be able to

1. Perform the experiment your lab team designed.
2. Present and interpret the results of your experiment.

Before you do the experiment, be sure that everyone on your lab team understands the rules of the simulation.

Be thorough in collecting data. Don't just write down numbers; record what they mean as well. Don't rely on your memory for information that you will need when reporting on your experiment later! If you have any questions, doubts, or problems during the investigation, be sure to write them down, too.

Results

Before you begin to prepare your results for presentation, decide on the best format to use. Remember, you want to give the reader a clear, concise picture of what your experiment showed. Refer to the data presentation section of Appendix A (Tools for Scientific Inquiry) for help. If you are drawing graphs, use graph paper. Complete your tables and/or graphs before attempting to interpret your results.

Write a few sentences *describing* the results. Don't explain why you got these results or draw conclusions yet.

Discussion

Look back at the hypothesis you posed in this experiment. Look at the graphs or tables of your data. Do your results support your hypothesis or prove it false? Explain your answer, using your data for support.

Did your results correspond to the prediction you made? If not, explain how your results are different from your expectations and why this might have occurred.

Describe how your data are supported by information from other sources (for example, textbooks or other lab teams working on a similar problem).

If you had any problems with the procedure or had questionable results, explain how they might have influenced your conclusion.

If you had an opportunity to repeat and extend this experiment to make your results more convincing, what would you do?

Summarize the conclusion you have drawn from your results.

Questions for Review

1. Putting beans in a pan of sand is only one way to simulate foraging. Use your imagination and think of other simulation systems that might be used.

2. Give some examples of how foragers or predators recognize their prey as prey.

3. From your experience with the simulation, do you think that foragers are more efficient if they work in cooperative groups or if they work solo? Explain your answer.

4. If a forager has a choice of prey items available, what factors might influence its prey selection?

5. The ability of the prey to reproduce has a significant effect on the survival of the population. In nature, what factors might affect the ability of the members of a prey population to reproduce?

6. Give some examples from nature of different types of "refuges" used by prey species.

Acknowledgment

The idea of foraging for beans in pans of sand came from Hans Landel, who was a graduate student at Purdue University at the time.

Tools for Scientific Inquiry

Introduction

In addition to using the particular method of approaching and solving problems that is known as scientific inquiry, scientists also use standard methods of collecting and presenting data.

Scientists use the metric system to take measurements. This appendix reviews metric measurements and techniques for using some laboratory measuring devices.

When scientists communicate the results of their experiments to others, they present numerical data in a format that is clear and concise. This appendix also explains how to present and interpret data.

Outline

Exercise A.1: Scientific Notation

Exercise A.2: Making Measurements

> Activity A: Measuring Length

> Activity B: Measuring Mass

> Activity C: Measuring Volume

> Activity D: Measuring Temperature

Exercise A.3: Data Presentation

> Activity A: Tables

> Activity B: Graphs

> Activity C: Graphing Practice

Exercise A.4: Interpreting Information on a Graph

EXERCISE A.1
Scientific Notation

This exercise explains how to

1. Convert numbers that are expressed in scientific notation to decimals.
2. Express decimal numbers in scientific notation.
3. Multiply and divide numbers that are expressed in scientific notation.

Scientific notation is a numbering system that can easily handle the very small or very large numbers that are sometimes used in scientific investigations. The form of a number expressed in scientific notation is $a \times 10^n$, where a is a whole number or a decimal between 1 and 10 and n (the exponent) is a whole number.

For numbers greater than 1, the exponent is positive and indicates the number of decimal places to the right of the decimal point in a. Examples:

$1.2 \times 10^5 = 120{,}000$

$4 \times 10^{12} = 4{,}000{,}000{,}000{,}000$

$4.65 \times 10^6 = 4{,}650{,}000$

For numbers less than 1, the exponent is negative and indicates the number of decimal places to the left of the decimal point in a. Examples:

$3 \times 10^{-4} = 0.0003$

$1.7 \times 10^{-9} = 0.0000000017$

$1.0 \times 10^{-1} = 0.1$

$5.74 \times 10^{-6} = 0.00000574$

For practice, write the following numbers in decimals:

$2.3 \times 10^{-3} =$	$6.43 \times 10^{-6} =$
$4.64 \times 10^2 =$	$1.66 \times 10^4 =$
$8.9 \times 10^{-7} =$	$4.3 \times 10^6 =$
$1.0 \times 10^5 =$	$7.2 \times 10^{-5} =$

Write the following decimal numbers in scientific notation:

$0.0032 =$	$23{,}000 =$
$1{,}000{,}000 =$	$0.45 =$
$0.0000061 =$	$71{,}000{,}000{,}000 =$
$0.01 =$	$0.000000009 =$

Multiplication

When two numbers in scientific notation are multiplied, the numbers are multiplied but the exponents are added together. Examples:

$(2 \times 10^3)(1.5 \times 10^2) = 3 \times 10^5$ (check by multiplying $2{,}000 \times 150$)

$(2 \times 10^{-3})(1.5 \times 10^2) = 3 \times 10^{-1}$ ($0.002 \times 150 = 0.3$)

$(2 \times 10^{-3})(1.5 \times 10^{-2}) = 3 \times 10^{-5}$ ($0.002 \times 0.015 = 0.00003$)

For practice, do the following problems:

$(3 \times 10^3)(6 \times 10^2) =$

$(7 \times 10^{-2})(9 \times 10^4) =$

$(1.2 \times 10^{-6})(3.0 \times 10^{-2}) =$

$(1.0 \times 10^{-8})(4.67 \times 10^9) =$

Division

When two numbers expressed in scientific notation are divided, the exponent of the divisor is subtracted from the exponent of the dividend.

$$\frac{2 \times 10^6}{2 \times 10^4} = 1 \times 10^2 \qquad \text{Check:}\ \frac{2{,}000{,}000}{20{,}000} = \frac{200}{2} = 100$$

$$\frac{7.2 \times 10^4}{3 \times 10^2} = 2.4 \times 10^2 \qquad \text{Check:} \frac{72,000}{300} = 240$$

$$\frac{7.2 \times 10^4}{3 \times 10^{-2}} = 2.4 \times 10^6 \qquad \text{Check:} \frac{72,000}{0.03} = 2,400,000$$

$$\frac{7.2 \times 10^{-4}}{3 \times 10^{-2}} = 2.4 \times 10^{-2} \qquad \text{Check:} \frac{0.00072}{0.03} = 0.024$$

$$\frac{7.2 \times 10^{-4}}{3 \times 10^2} = 2.4 \times 10^{-6} \qquad \text{Check:} \frac{0.00072}{300} = 0.0000024$$

For practice, do the following problems:

$$\frac{1.2 \times 10^{-4}}{3 \times 10^{-3}} = \qquad\qquad =$$

$$\frac{4.8 \times 10^5}{3 \times 10^{-7}} =$$

$$\frac{2.6 \times 10^2}{2 \times 10^7} =$$

$$\frac{2.1 \times 10^{-6}}{1 \times 10^2} =$$

EXERCISE A.2
Making Measurements

This exercise explains how to

1. Convert metric units of length, mass, and volume.
2. Use a balance.
3. Use a pipet.
4. Use a graduated cylinder.
5. Convert temperatures expressed in degrees Fahrenheit to degrees Celsius and vice versa.

Many of the labs or investigations you do in this course require you to make accurate measurements using the metric system. In this exercise you will review metric units and learn to use the measuring devices that you are most likely to encounter in the course.

Activity A: Measuring Length

The **meter** (m) is the basic unit of linear measurement in the metric system. As a comparison to American Standard units, a meter is a little longer than a yard. Longer distances are measured in **kilometers** (km). One kilometer equals 1,000 m and is approximately six-tenths of a mile. The meter is subdivided into **centimeters** (cm) and **millimeters** (mm). One centimeter is one one-hundredth of a meter (1 cm = 0.01 m). One millimeter is one one-thousandth of a meter (1 mm = 0.001 m). **Micrometers** (µm), which are also called microns, and **nanometers** (nm) are smaller subdivisions that are used for measurements of microscopic objects (see Table A.1).

Table A.1
Metric Units for Linear Measurement

Unit	Meters
Kilometer (km)	1×10^3
Meter (m)	1
Centimeter (cm)	1×10^{-2}
Millimeter (mm)	1×10^{-3}
Micrometer (also called micron; µm)	1×10^{-6}
Nanometer	1×10^{-9}

The metric side of the ruler at your station is numbered in centimeters, and each centimeter is divided into 10 mm. Compare the metric side of the ruler with the American Standard side. Approximately how many cm are there in one inch?_____ How many mm in one inch?_____ How many inches in one cm?_____

Metric units can be converted by using Table A.1 to determine the conversion factor. The equation is set up so that similar units are crossed out. For example, to convert mm to cm, divide the number of m per mm (given in Table A.1) by the number of m per cm. The meter units are crossed out, leaving a factor that relates cm to mm.

$$\frac{1 \times 10^{-3} \, \cancel{m}}{1 \, mm} \times \frac{1 \, cm}{1 \times 10^{-2} \, \cancel{m}} = \frac{1 \times 10^{-1} \, cm}{1 \, mm} \qquad \text{(conversion factor)}$$

Then, to convert some number (such as 25 mm), to cm:

$$25 \, \cancel{mm} \, \frac{(1 \times 10^{-1} \, cm)}{1 \, \cancel{mm}} = 25 \times 10^{-1} \, cm = 2.5 \, cm$$

(Note that the mm units cross out, leaving the answer in cm.)

This example illustrates how to convert 3.4 cm to μm:

$$\frac{1 \times 10^{-2}\,m}{1\,cm} \times \frac{1\,\mu m}{1 \times 10^{-6}\,m} = \frac{1 \times 10^4\,\mu m}{1\,cm} \qquad \text{(conversion factor)}$$

$$3.4\,cm\,\frac{(1 \times 10^4\,\mu m)}{1\,cm} = 3.4 \times 10^4\,\mu m = 34{,}000\,\mu m$$

Practice using metric units by doing the following conversions:

1500 m = _____ km   350 nm = _____ μm

1.2 cm = _____ mm   10 km = _____ m

100 cm = _____ m   175 mm = _____ cm

600 nm = _____ μm   175 mm = _____ m

Convert the measurements in Table A.2 as indicated.

Table A.2
Relative Size (in meters)

Thickness of plasma membrane	7×10^{-9} m to 10×10^{-9} m = _____ nm
Length of *E. coli* bacteria	1×10^{-6} m = _____ μm
Diameter of a typical animal cell	1×10^{-5} m to 3×10^{-5} m = _____ μm
Diameter of a human egg	1×10^{-4} m = _____ μm
Diameter of a frog egg	1×10^{-3} m = _____ μm
Diameter of a human eyeball	2.5×10^{-2} m = _____ cm
Length of a human esophagus	2.4×10^{-1} m = _____ cm
Height of an average human female	1.6 m = _____ cm
Length of a human small intestine	6.0 m = _____ cm

Activity B: Measuring Mass

The **gram** (g) is the basic unit for measuring mass in the metric system. Larger amounts are measured in **kilograms** (kg). One kilogram = 1,000 g. As a comparison to American Standard units, a kilogram is approximately 2.2 pounds. It takes approximately 28 g to make an ounce and 454 g to make a pound. Like the meter, the gram is also divided into smaller units (see Table A.3).

Table A.3
Metric Units for Measuring Mass

Unit	Grams
Kilogram (kg)	1×10^3
Gram (g)	1
Milligram (mg)	1×10^{-3}
Microgram (μg)	1×10^{-6}
Nanogram (ng)	1×10^{-9}

Converting metric mass units can be done using the same conversion factors that you used for length units. For example, to convert 35 milligrams to micrograms:

$$\frac{1 \times 10^{-3}\,\cancel{g}}{1\,mg} \times \frac{1\,\mu g}{1 \times 10^{-6}\,\cancel{g}} = \frac{1 \times 10^3\,\mu g}{1\,mg}$$

$$35\,\cancel{mg}\ \frac{(1 \times 10^3\,\mu g)}{1\,\cancel{mg}} = 35 \times 10^3\,g = 3.5 \times 10^4\,mg$$

Practice using metric mass units by doing the following conversions:

60 mg = _____ g   2 μg = _____ ng

240 g = _____ kg   0.4 kg = _____ g

1.7 mg = _____ μg   3 ng = _____ μg

The following procedures describe the two types of instruments that are most commonly used in biology laboratories to determine weight, the triple-beam balance and the electronic balance. If your laboratory is equipped with another type of balance, your instructor will explain how to use it.

With both the triple-beam and electronic balances, a plastic or aluminum "boat" or a thin paper is used to hold the sample you are weighing. Therefore, you must set the balance to compensate for the weight of the boat or paper before weighing the sample. This is called **taring** the balance. To tare the balance, the boat or paper is placed on the balance and its weight is set at 0. Thus the weight of the container is not added to the weight of the sample.

Several objects are provided at your work station, along with a weighing boat. You will practice using the balances by weighing these objects. A weighing boat would not normally be required for the type of objects used in this exercise, but you should practice using it so you will learn how to tare the balance. In other laboratories you will be weighing samples that do require a weighing boat.

The Triple-Beam Balance

A triple-beam balance is illustrated in Figure A.1.

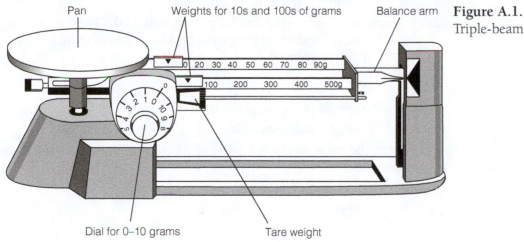

Pan   Weights for 10s and 100s of grams   Balance arm

Dial for 0–10 grams   Tare weight

Figure A.1.
Triple-beam balance.

Procedure

Taring the Balance

1. Be sure all the weights are set at 0.
2. Place the weighing boat on the balance pan.
3. Adjust the tare weight right or left until the arm balances at the 0 mark. (You should not have to move the weight more than a few mm.)

Weighing the Sample

1. Place the sample in the weighing boat.
2. Slide the weight on the middle beam to the right so that it fits in the first (100-g) notch.
   a. If the arm drops below the 0 mark, the item weighs less than 100 g. Slide the weight back to the left and go to step 3.
   b. If the balance arm remains above the 0 mark, the object weighs more than 100 g. Slide the weight over to the 200-g notch. If the arm then drops below the 0 mark, the object weighs between 100 and 200 g. Move the weight back to the 100-g notch. If the balance arm is still above the 0 mark, go to the 300-g notch, and so on.
3. Follow the same procedure with the back beam, which has unit markings in tens of grams.
4. The dial has unit markings in single grams and one-tenth of a gram. Turn the dial until the arm balances at the 0 mark.
5. Record the weight of the object.

The Electronic Balance

Figure A.2 shows a typical electronic balance. There are many different models of this kind of balance, so it is unlikely that the balances in your laboratory look exactly like this. Your instructor will explain any additional features on your balance model that you should be able to use.

The balances should be turned on, warmed up, and ready for use.

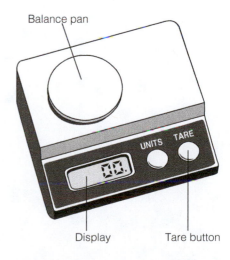

Balance pan

Display   Tare button

Figure A.2.
Electronic balance.

Procedure

Taring the Balance
1. Set the weighing boat on the balance pan.
2. Press the tare button. The display should read 0.00. (The readout may be different, depending on the model of balance you're using.) The last digit may drift somewhat.

Weighing the Sample
1. Place the sample in the weighing boat.
2. Read the weight from the display.
3. Record the name of the sample and its weight below. Remember to record the units.

Practice by weighing several objects.

Object	Weight
_____	_____
_____	_____
_____	_____
_____	_____

Activity C: Measuring Volume

The **liter** (L) is the basic unit of volume of fluids in the metric system. As a comparison to American Standard units, a liter is slightly larger than a quart. The liter is subdivided into **milliliters** (mL). One liter = 1,000 mL. Milliliters can be further subdivided into **microliters** (µL). One milliliter = 1,000 µL. (See Table A.4.)

Table A.4
Metric Units for Fluid Volume

Unit	Liters
Liter (L)	1
Milliliter (mL)	1×10^{-3}
Microliter (µL)	1×10^{-6}

Practice using metric units by doing the following conversions.

4 mL = _____ L   10 L = _____ mL

300 mL = _____ µL   37 mL = _____ L

0.2 L = _____ mL   750 µL = _____ mL

Pipets

To measure a small volume (10 mL or less), you will use 1-, 5-, or 10-mL pipets.

Two commonly used types of pipets are blowout and delivery pipets. With a blowout pipet, the volume is measured all the way into the tip, and accurate measurement is obtained by expelling every last drop of liquid. In a delivery pipet, the tip is unmeasured "dead space." Liquid remaining beyond the last marking should be discarded. If both types are used in your lab, get a pipet of each type and note the differences you see.

Look at the tops of the pipets. The largest and smallest volumes you can measure are given there. For example, it may say "5 mL in 1/10." This means that the pipet measures 5 mL from the 0 line to the 5-mL line (for a delivery pipet) or the tip (for a blowout pipet). Each line in between the mL marks represents 1/10 mL.

Liquid is drawn into the pipet using a rubber bulb or some other type of filling device. In this laboratory manual we will refer to a filling device called a pipet pump, or pi-pump. Use of the pi-pump will be demonstrated in the following procedure. If your lab uses a different device, your instructor will explain how to use it.

⚠ **Never pipet by mouth. Mouth pipeting is strictly prohibited.**

Pi-pumps come in different sizes. Blue pi-pumps fit 1-mL pipets. Green pi-pumps fit 5- and 10-mL pipets.

Procedure for Blowout Pipets

1. Get a 5-mL blowout pipet and a green pi-pump.
2. Attach the pi-pump to the pipet.

⚠ **Hold the pipet at the top and insert it firmly but gently into the pi-pump. Never force the pi-pump onto the pipet.**

3. Put the pipet into the flask of red solution on your lab table. Hold the barrel of the pi-pump with four fingers, and turn the wheel clockwise with your thumb (see Figure A.3). This will suck liquid into the pipet.
4. Fill the pipet to the 0 mark.

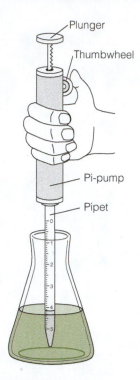

Figure A.3.
Using a pi-pump to fill a pipet.

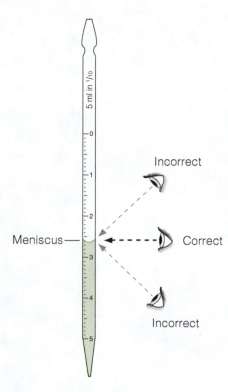

Figure A.4.
Reading the meniscus in a pipet.

5. In Figure A.4, you can see that the liquid level may be read at different angles. To read the level correctly, your eye should be directly in line with the **meniscus**, which is the crescent formed at the surface of the liquid. The *bottom* of the meniscus marks the correct volume.

6. Move the pipet to the waste container. Turn the thumbwheel counterclockwise to release the liquid. Depress the plunger to expel the last bit.

    What volume of liquid have you measured into the container?

7. Refill the pipet with red solution to the 0 mark.

8. Move the pipet to the waste container. Turn the thumbwheel slowly to release liquid down to the 3 mark. Remember to read the bottom of the meniscus.

    What volume of liquid have you measured?

9. Discard the remaining liquid in the waste container.

> ✴ To avoid contamination, you should never return liquid to the container you got it from.

Procedure for Delivery Pipets

1. Get a 5-mL delivery pipet and a green pi-pump.
2. Attach the pi-pump to the pipet.

> ⚠ Hold the pipet at the top and insert it firmly but gently into the pi-pump. Never force the pi-pump onto the pipet.

3. Put the pipet into the flask of red solution. Hold the barrel of the pi-pump with four fingers and turn the wheel clockwise with your thumb (see Figure A.3). This will suck liquid into the pipet.

4. Fill the pipet to the 0 mark.

5. In Figure A.4, you can see that the liquid level may be read at different angles. To read the level correctly, your eye should be directly in line with the **meniscus.** The bottom of the meniscus marks the correct volume.

6. Move the pipet to the waste container. Turn the thumbwheel slowly to release the liquid. Stop when the bottom of the meniscus reaches the 5 mL mark.

What volume of liquid have you measured into the container?

7. Liquid remaining in the pipet should be discarded in the waste container.

✳   To avoid contamination, never return liquid to the container you got it from. To expel all the remaining liquid from the pipet, push down the plunger.

---

8. Refill the pipet with red solution to the 0 mark.
9. Move the pipet to the waste container. Turn the thumbwheel slowly to release liquid down to the 3 mark. Remember to read the bottom of the meniscus.

What volume of liquid have you measured?

Remember to discard the remaining liquid in the waste container.

Practice measuring with pipets as directed by your instructor.

Graduated Cylinders

Graduated cylinders (grad cylinders) are used to measure larger amounts of fluid, although you may use a 10-mL graduated cylinder in place of a 10-mL pipet. Before you use a graduated cylinder, check to see what measurements its markings represent. For example, a typical 100-mL graduated cylinder is marked at 1-mL increments, with total volume labeled every 10 mL.

Procedure

1. Carefully pour the red solution from its container into the 100-mL grad cylinder, measuring a volume of 35 mL.

✳   If you are using a glass cylinder, remember to read the bottom of the meniscus. No meniscus is formed in plastic labware.

---

2. Add the solution from the waste container to the grad cylinder. What is the total volume now?

3. When you are all finished with Activity C, pour the red solution from the grad cylinder back into the flask for the next team of students to use. (Since you are just practicing measurement techniques, contamination is not a concern.)

Beakers and Flasks

Most beakers and flasks have volume markings on them. However, they are not very accurate. If a procedure tells you to use 100 mL of a solution, you should measure it with a graduated cylinder. If the procedure says to use *approximately* 100 mL, the markings on a beaker or flask are close enough.

Activity D: Measuring Temperature

Scientific temperature measurements are made using the Celsius (which used to be called centigrade, "100 degree") scale, rather than the Fahrenheit scale. Conversions are made using the following formulas:

$$°C = (°F - 32)\frac{5}{9}$$

$$°F = (°C \times \frac{9}{5}) + 32$$

Table A.5 shows several comparisons between Celsius and Fahrenheit temperatures.

Table A.5
Comparison of Temperatures

	°F	°C
A very cold day	10	−12.2
Freezing point of water	32	0
A cool day	50	10
Room temperature	72	22.2
Human body temperature	98.2	36.8
Boiling point of water	212	100

Practice converting temperatures with the following problems.

145°F = _____ °C   28°C = _____ °F

16°F = _____ °C   72°C = _____ °F

68°F = _____ °C   6°C = _____ °F

EXERCISE A.3
Data Presentation

This exercise explains how to
1. Distinguish between discrete and continuous variables.
2. Construct a line graph.
3. Construct a bar graph.
4. Choose the best method for presenting your data.

Activity A: Tables

A student team performed the experiment that was discussed in Lab Topic 1. They tested the pulse and blood pressure of basketball players and nonathletes to compare cardiovascular fitness. They recorded the following data:

	Nonathletes							Basketball Players					
	Resting Pulse			After Exercise				Resting Pulse			After Exercise		
	Trial			Trial				Trial			Trial		
Subject	1	2	3	1	2	3	Subject	1	2	3	1	2	3
1	72	68	71	145	152	139	1	67	71	70	136	133	134
2	65	63	72	142	144	158	2	73	71	70	141	144	142
3	63	68	70	140	147	144	3	72	74	73	152	146	149
4	70	72	72	133	134	145	4	75	70	72	156	151	151
5	75	76	77	149	152	153	5	78	72	76	156	150	155
6	75	75	71	154	148	147	6	74	75	75	149	146	146
7	71	68	73	142	145	150	7	68	69	69	132	140	136
8	68	70	66	135	137	135	8	70	71	70	151	148	146
9	78	75	80	160	155	153	9	73	77	76	138	152	147
10	73	75	74	142	146	140	10	72	68	64	153	155	155

If the data were presented to readers like this, they would see just lists of numbers and would have difficulty discovering any meaning in them. This is called raw data. It shows the data the team collected without any kind of summarization. Since the students had each subject perform the test three times, the data for each subject can be averaged. The other raw data sets obtained in the experiment would be treated in the same way.

Table. Average Pulse Rate for Each Subject
(Average of 3 trials for each subject; pulse taken before and after 5-min step test)

	Nonathletes			Basketball Players	
	Resting Pulse	After Exercise		Resting Pulse	After Exercise
Subject	Average	Average	Subject	Average	Average
1	70	145	1	70	134
2	67	148	2	70	142
3	67	144	3	73	149
4	71	139	4	72	151
5	76	151	5	76	155
6	74	150	6	75	146
7	71	146	7	69	136
8	68	136	8	70	146
9	78	156	9	76	147
10	74	143	10	68	155

These rough data tables are still rather unwieldy and hard to interpret. A summary table could be used to convey the overall averages for each part of the experiment. For example:

Table. Overall Averages of Pulse Rate
(10 subjects in each group; 3 trials for each subject;
pulse taken before and after 5-min step test)

Pulse Rate (beats/min)		
	Before Exercise	After Exercise
Nonathletes	71.6	145.8
Basketball players	71.9	146.1

Notice that the table has a title above it that describes its contents, including the experimental conditions and the number of subjects and replications that were used to calculate the averages. In the table itself, the units of the dependent variable (pulse rate) are given and the independent variable (nonathletes and basketball players) is written on the left side of the table.

Tables should be used to present results that have relatively few data points. Tables are also useful to display several dependent variables at the same time. For example, average pulse rate before and after exercise, average blood pressure before and after exercise, and recovery time could all be put in one table.

Activity B: Graphs

Numerical results of an experiment are often presented in a graph rather than a table. A graph is literally a picture of the results, so a graph can often be more easily interpreted than a table. Generally, the independent variable is graphed on the *x* axis (horizontal axis) and the dependent variable is graphed on the *y* axis (vertical axis). In looking at a graph, then, the effect that the independent variable has on the dependent variable can be determined. "Dry mix" may help you remember which variable goes on which axis. The **d**ependent variable is sometimes called the **r**esponding variable and it goes on the **y** axis. **M**anipulated variable is a term sometimes used for the **i**ndependent variable, and it goes on the **x** axis.

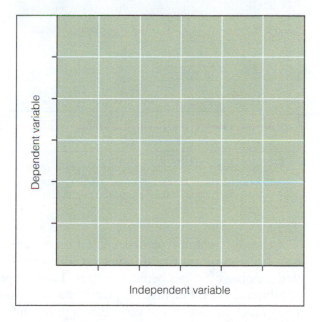

Figure A.5.
Graph construction.

When you are drawing a graph, keep in mind that your objective is to show the data in the clearest, most readable form possible. In order to achieve this, you should observe the following rules:

- Use graph paper to plot the values accurately.
- Plot the independent variable on the *x* axis and the dependent variable on the *y* axis. For example, if you are graphing the effect of the amount of fertilizer on peanut weight, the amount of fertilizer is plotted on the *x* axis and peanut weight is plotted on the *y* axis.
- Label each axis with the name of the variable and specify the units used to measure it. For example, the *x* axis might be labeled "Fertilizer

applied (g/100 m²)" and the y axis might be labeled "Weight of peanuts per plant (grams)."

- The intervals labeled on each axis should be appropriate for the range of data so that most of the area of the graph can be used. For example, if the highest data point is 47, the highest value labeled on the axis might be 50. If you labeled intervals on up to 100, there would be a large unused area of the graph.

- The intervals that are labeled on the graph should be evenly spaced. For example, if the values range from 0 to 50, you might label the axis at 0, 10, 20, 30, 40, and 50. It would be confusing to have labels that correspond to the actual data points (for example, 2, 17, 24, 30, 42, and 47).

- The graph should have a title that, like the title of a table, describes the experimental conditions that produced the data.

Figure A.6 illustrates a well-executed graph.

Figure A.6.
Graph of peanut weight vs. amount of fertilizer applied.

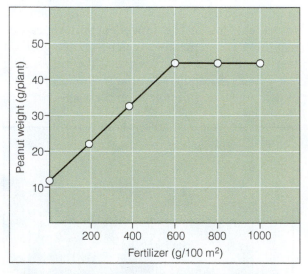

Figure 1. Weight of peanuts produced per plant when amount of fertilizer applied is varied. (Average seed weight per plant in 100 m² plots, 400 plants/plot.)

The most commonly used forms of graphs are line graphs and bar graphs. The choice of graph type depends on the nature of the independent variable being graphed. **Continuous variables** are those that have an unlimited number of values between points. **Line graphs** are used to represent continuous data. For instance, time is a continuous variable over which things such as growth will vary. Although the units on the axis can be minutes, hours, days, months, or even years, values can be placed in between any two values. Amount of fertilizer can also be a continuous variable. Although the intervals labeled on the x axis are 0, 200, 400, 600, 800, and 1,000 g/100 m², many other values can be listed between each two intervals.

In a line graph, data are plotted as separate points on the axes, and the points are connected to each other. Notice in Figure A.7 that when there is more than one set of data on a graph, it is necessary to provide a key indicating which line corresponds to which data set.

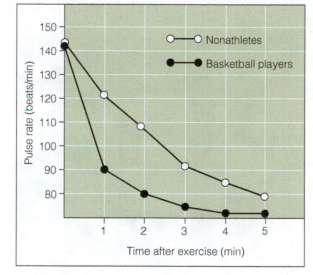

Figure A.7.
Line graph representing two related sets of data.

Figure 1. Recovery rate of basketball players and nonathletes after performing a step test for 5 minutes. (Average of 10 subjects; each subject performed the test 3 times.)

Discrete variables, on the other hand, have a limited number of possible values, and no values can fall between them. For example, the type of fertilizer is a discrete variable: There are a certain number of types which are distinct from each other. If fertilizer type is the independent variable displayed on the *x* axis, there is no continuity between the values.

Bar graphs, as shown in Figure A.8, are used to display discrete data.

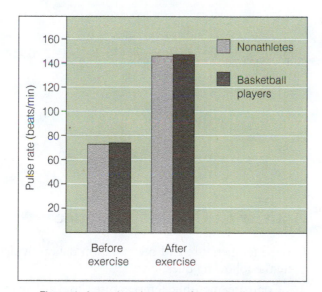

Figure A.8.
Example of bar graph.

Figure 1. Average pulse rates of basketball players and nonathletes before and after performing a step test for 5 minutes. (Average of 10 subjects; each subject performed the test 3 times.)

In this example, before- and after-exercise data are discrete: There is no possibility of intermediate values. The subjects used (basketball players and nonathletes) also are a discrete variable (a person belongs to one group or the other).

This graph could also have been constructed as shown in Figure A.9.

Figure A.9.
Alternative method of presenting data in Figure A.7.

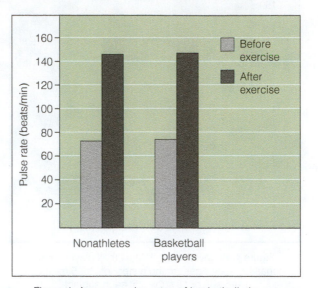

Figure 1. Average pulse rates of basketball players and nonathletes before and after performing a step test for 5 minutes. (Average of 10 subjects; each subject performed the test 3 times.)

What is the difference between the two graphs?

Explain why the first way would be better to convey the results of the experiment.

Activity C: Graphing Practice

Use the temperature and precipitation data provided in Table A.6 to complete the following questions:

1. Compare monthly temperatures in Fairbanks with temperatures in San Salvador.

    Can data for both cities be plotted on the same graph?

    What will go on the *x* axis?

Table A.6

Average Monthly High Temperature and Precipitation for Four Cities
(T = temperature in °C; P = precipitation in cm)

		Jan.	Feb.	Mar.	Apr.	May	June	July	Aug.	Sept.	Oct.	Nov.	Dec.
Fairbanks, Alaska	T	−19	−12	−5	6	15	22	22	19	12	2	−11	−17
	P	2.3	1.3	1.8	0.8	1.5	3.3	4.8	5.3	3.3	2.0	1.8	1.5
San Francisco, California	T	13	15	16	17	17	19	18	18	21	20	17	14
	P	11.9	9.7	7.9	3.8	1.8	0.3	0	0	0.8	2.5	6.4	11.2
San Salvador, El Salvador	T	32	33	34	34	33	31	32	32	31	31	31	32
	P	0.8	0.5	1.0	4.3	19.6	32.8	29.2	29.7	30.7	24.1	4.1	1.0
Indianapolis, Indiana	T	2	4	9	16	22	28	30	29	25	18	10	4
	P	7.6	6.9	10.2	9.1	9.9	10.2	9.9	8.4	8.1	7.1	8.4	7.6

Source: Pearce, E. A., and G. Smith. Adapted from *The Times Books World Weather Guide.* New York: Times Books, 1990.

How should the *x* axis be labeled?

What should go on the *y* axis?

What is the range of values on the *y* axis?

How should the *y* axis be labeled?

What type of graph should be used?

2. Compare the average September temperature for Fairbanks, San Francisco, San Salvador, and Indianapolis.
    Can data for all four cities be plotted on the same graph?

What will go on the *x* axis?

How should the *x* axis be labeled?

What should go on the *y* axis?

What is the range of values on the *y* axis?

How should the *y* axis be labeled?

What type of graph should be used?

3. Graph the temperature and precipitation data for San Francisco. Can both sets of data be plotted on the same graph?

What will go on the *x* axis?

How should the *x* axis be labeled?

What should go on the *y* axis?

What is the range of values on the temperature axis?

How should this axis be labeled?

What is the range of values on the precipitation axis?

How should this axis be labeled?

What type of graph should be used?

E X E R C I S E   A . 4
Interpreting Information on a Graph

This exercise explains how to
1.   Interpret graphs.

Once you understand how graphs are constructed, it is easier to get information from the graphs in your textbook as well as to interpret the results you obtain from laboratory experiments. For the graphs below, write a sentence or two describing what each graph shows, and answer the questions.

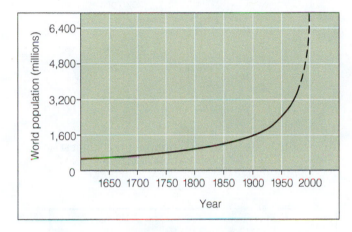

Figure A.10.
Change in world population from 1650 to 2000.

Describe what the graph shows.

What was the world's population in 1900?

Predict the world's population in 2000.

Figure A.11.
Rate of an enzymatic reaction at
different temperatures.

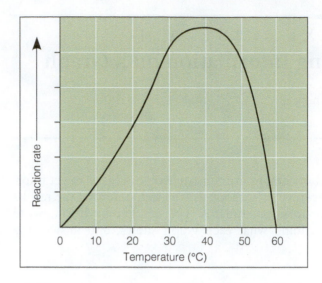

Describe what the graph shows.

At what temperature is reaction rate the highest?

Figure A.12.
Change in range of average
monthly temperature as
latitude changes.

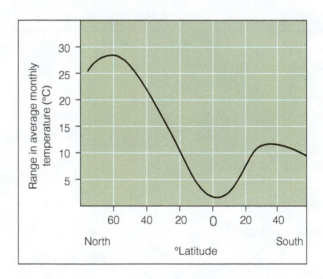

Describe what the graph shows.

At what latitude does the least variation in temperature occur?

Miami is at approximately 26° N latitude. From the information on the graph, what is the range in mean monthly temperature there?

Minneapolis is at approximately 45° N latitude. From the information on the graph, what is the range in mean monthly temperature there?

Sydney, Australia is at approximately 33° S latitude. From the information on the graph, what is the range in mean monthly temperature there?

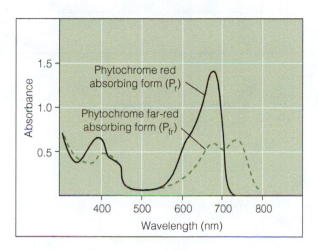

Figure A.13.
Absorption of light by the pigments P_r phytochrome and P_{fr} phytochrome.

Describe what the graph shows.

At what wavelengths does P_r phytochrome absorb the most light?

At what wavelengths does P_{fr} phytochrome absorb the most light?

Figure A.14.
Population growth of *Paramecium aurelia* and *Paramecium caudatum* when grown separately and together.

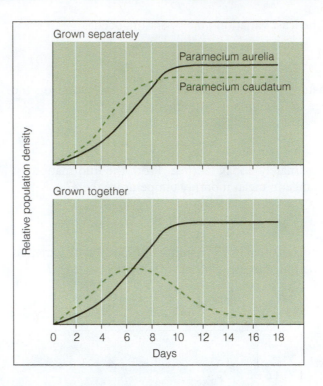

Describe what the graphs show.

On what day does *Paramecium aurelia* reach its maximum population density?

Does *Paramecium caudatum* do better when it is grown alone or when it is grown in a mixture with *Paramecium aurelia*?

Questions for Review

1. Convert the following numbers to scientific notation.

   0.0032 =   7,310,000 =

   457,200,000 =   689 =

   0.67 =   0.000483 =

2. Convert the following numbers to decimals.

   7.32×10^{-4} = 1.34×10^{8} =

   3.43×10^{3} = 7.55×10^{-6} =

   6.2×10^{5} = 1.8×10^{-2} =

3. Perform the following calculations.

   $$\frac{3.4 \times 10^{-2}}{1.7 \times 10^{-5}} = \qquad \frac{1.0 \times 10^{5}}{2.0 \times 10^{3}} = \qquad =$$

   $$\frac{8.6 \times 10^{-3}}{2.0 \times 10^{2}} = \qquad \frac{9.9 \times 10^{6}}{3.3 \times 10^{-4}} =$$

   $(3.5 \times 10^{-2})(2.0 \times 10^{-5})$ =

   $(7.36 \times 10^{3})(1.0 \times 10^{-6})$ =

   $(1.6 \times 10^{4})(3.0 \times 10^{2})$ =

   $(1.11 \times 10^{-5})(6.0 \times 10^{9})$ =

4. Convert the following measures.

2.8 mm =	nm	4.67 m =	μm
1.3 nm =	μm	67 cm =	m
12 μg =	ng	1.6 g =	kg
300 μg =	g	250 mg =	μg
83 mL =	L	250 mL =	L
175 μL =	mL	0.5 L =	mL
75 °F =	°C	50 °C =	°F

In the following questions you will be constructing graphs without plotting data. By practicing how to construct graphs, you will learn how to graph your own data in later labs.

5. A team of students hypothesizes that the amount of alcohol produced in fermentation depends on the amount of sugar supplied to the yeast. They want to use 5, 10, 15, 20, 25, and 30% sugar solutions. They propose to run each experiment at 40°C with 5 mL of yeast.

   What type of graph is appropriate for presenting these data? Explain why.

Sketch the axes of a graph that would present these data. Mark the intervals on the x axis and label both axes completely. Write a title for the graph.

6. Having learned that the optimum sugar concentration is 25%, the students next decide to investigate whether different strains of yeast produce different amounts of alcohol. If you were going to graph the data from this experiment, what type of graph would be used? Explain why.

Sketch and label the axes for this graph and write a title.

7. A team of students wants to study the effect of temperature on bacterial growth. They put the dishes in different places: an incubator (37°C), a lab room (21°C), a refrigerator (10°C), and a freezer (0°C). Bacterial growth is measured by estimating the percentage of each dish that is covered by bacteria at the end of a 3-day growth period.

   What type of graph would be used to present these data? Explain why.

   Sketch the axes below. Mark the intervals on the x axis, and label both axes completely. Write a title for the graph.

8. A team of scientists is testing a new drug, XYZ, on AIDS patients. The scientists monitor patients in the study for symptoms of 12 different diseases. What would be the best way for them to present these data? Explain why.

9. A group of students decides to investigate the loss of chlorophyll in autumn leaves. They collect green leaves and leaves that have turned color from sugar maple, sweet gum, beech, and aspen trees. Each leaf is subjected to an analysis to determine how many mg of chlorophyll is present.

   What type of graph would be most appropriate for presenting the results of this experiment? Explain why.

Sketch the axes for the graph below to show how you would present the data. Write a title for the graph.

Guide to Writing a Scientific Report

Purpose of the Lab Report

No matter what you are writing, you should always know who your audience is so you can write in the way that best communicates with that audience. The "reader" referred to in this guide is a person who understands science but is not as knowledgeable about your particular subject as you are. Your job is to explain your investigation and its significance and to convince the reader that your conclusions are scientifically valid. Since you are writing your lab report for a grade, your instructor is also part of your audience. Your instructor wants to evaluate how well you can communicate your background knowledge and all the components of your investigation to a wider audience, so don't assume that your reader knows as much as your instructor about your investigation. Also keep in mind that writing a good paper depends more on *how well* you designed the experiment and can present and explain the results than on *what* results you obtained. Even if your experiment did not turn out as you expected, you can still write a good lab report.

A scientific report is organized so that each phase of the investigation is described and justified, from formulating a hypothesis through reaching a conclusion about the hypothesis. The exact format used may vary slightly from one source to another. For example, in the format presented here, the Discussion and Conclusion sections are separate; in some scientific journals a combined Discussion/Conclusion section is used.

This guide describes the purpose and content of each section of a report and provides a sample paper integrated throughout as an example. The sample report is the material printed in green. Margin notes alongside the sections of the sample report point out features in each part. Read this entire appendix before you begin writing your report. Be sure to read the checklist at the end of the appendix, too. After you've written your report, review the checklist to see whether you've fulfilled all of its criteria.

 The sample report is based on a fictitious experiment that was constructed solely to illustrate the composition of a scientific report. The conclusions stated are not based on scientific data.

Sections of the Report

Each section of your report appearing after the title should have a heading.

Title

The title is a statement of the problem you are investigating. It should contain key words that indicate what information the reader will find in the paper. Your hypothesis can be used as a basis for your title.

The title should be placed on a cover page that also includes your name, your instructor's name, and the date.

As an example of turning a hypothesis into a title, consider the following hypothesis: Dark-colored hair dye is more mutagenic to *Salmonella typhimurium* bacteria than is light-colored hair dye. You could reword this for your title:

The title may seem long, but it should give enough information about the investigation for readers to decide whether they are interested in reading the contents of the report.

The Effect of Hair Dye Color on Mutagenicity of *Salmonella typhimurium* Bacteria

Introduction

The Introduction tells the reader what your investigation was about. It provides information about the biological basis for your experiment and a very brief synopsis of your experimental design.

State your hypothesis clearly at the beginning of the Introduction. Follow it with an explanation of why you and your lab team thought this hypothesis was worth investigating.

Give enough background information and references to allow the reader to appreciate the significance of your experiment, including an explanation of any unfamiliar or technical terms. You should relate your investigation to the larger issues in this area of study.

Since you are not an expert, you must cite a reference to show the source of information for any statement you make about the biological basis for your investigation. This demonstrates that you have done your homework and learned about your subject before undertaking your experiment. Citing a reference means telling the reader where you got a particular piece of information. In the text of the report, an abbreviated citation form is used—for example: Campbell, Reece, Mitchell, and Taylor (2003). A section at the end of the report called Literature Cited gives details about the source of information so the reader can locate it. For your lab reports, your sources will include the lab manual and your textbook. Your instructor may provide additional sources or suggest that you visit the library to get additional information. You may also get information from your professors or other knowledgeable people.

The Introduction should also explain briefly how you intend to go about testing your hypothesis. Finally, state your predictions concerning the outcome of your experiment.

Introduction

This experiment was performed to investigate the following hypothesis: Dark hair dye is more mutagenic in *Salmonella typhimurium* bacteria than light hair dye.

The hypothesis is stated.

One of our team members had read in the newspaper (*Greenville News*, 2002) that using dark colors of hair dye increases a person's chances of developing certain kinds of cancer. It has been shown that normal cells can be changed to cancer cells by a change or mutation in the genetic material (DNA) of the cell. A chemical or other agent that changes the genetic material is called a mutagen. Mutagens that cause normal cells to become cancerous are called carcinogens (Campbell, Reece, Mitchel, and Taylor, 2003). In our experiment, we tested the mutagenicity, or ability to cause mutations, of three hair dyes: Basic Brunette (dark color), Bombshell Blonde (light color), and Raspberry Red (a red color in between the dark and light colors).

The author explains what led her to make this hypothesis.

Including this background information helps the reader understand the biological basis for the experiment and demonstrates to the instructor that the author knows the meanings of the terms used. Notice that the citations are set off in parentheses.

We used the Ames test, which is widely used to test whether a chemical causes mutations in *Salmonella typhimurium,* a species of bacteria. According to Campbell, Reece, Mitchell, and Taylor (2003), "In general, mutagens are carcinogens." If a chemical is mutagenic in bacteria, then it is studied further to determine whether it could be mutagenic or carcinogenic in animals or humans (Atlas, 1984).

A lab manual by Sigmon (2000) explained how to perform the Ames test using a strain of *Salmonella typhimurium* that cannot make the amino acid histidine as normal *Salmonella* can do. This strain of *Salmonella* bacteria, which is called his– (his minus), can't grow unless histidine is present in agar, the substance it grows on. The part of the bacterial DNA that codes for the ability to make histidine is sensitive to mutagens. A mutagen can cause a mutation that enables the bacteria to produce histidine (Figure B.1).

In this paragraph the author demonstrates that she understands the laboratory method used to test the hypothesis. The completeness of the explanation and Figures B.1 and B.2 will help the reader understand the author's interpretation of the results when it is presented in the Discussion section.

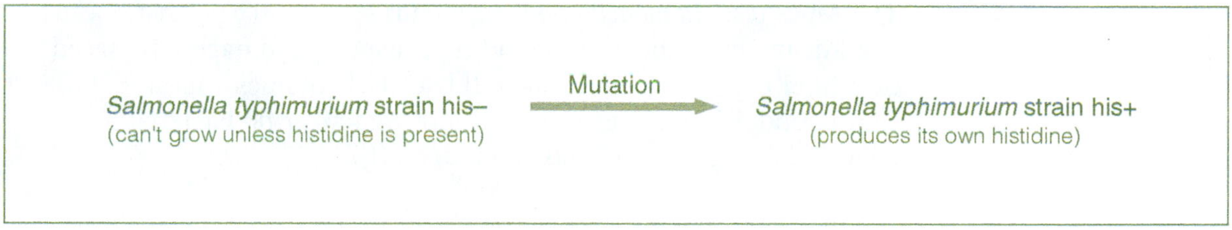

Mutation

Salmonella typhimurium strain his– → *Salmonella typhimurium* strain his+
(can't grow unless histidine is present)   (produces its own histidine)

Figure B.1.
Mutation in *Salmonella typhimurium*.

When a mutagen is present during the growth of the bacteria, then the bacteria gain the ability to produce histidine. The mutagenicity of a substance can be tested by growing the bacteria on agar that contains the suspected mutagen but doesn't contain histidine, as shown in Figure B.2.

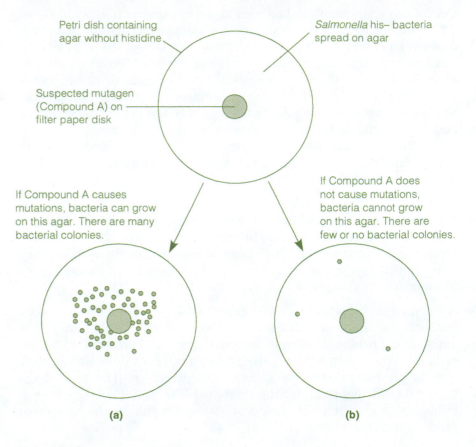

Figure B.2.
The Ames test. *Salmonella typhimurium* is spread on agar that doesn't contain any histidine. Compound A, a suspected mutagen, is placed on a filter paper disk on the agar. If bacterial colonies appear, then Compound A is a mutagen (a). If there are few or no bacterial colonies, Compound A is not a mutagen (b).

We used the Ames test for this investigation and applied brown, red, and blonde hair dyes to the filter paper circles. The growth of the his– strain of *Salmonella typhimurium* on the dishes indicated the ability of these dyes to cause mutations.

Predictions (if . . . then statements) made.

If our hypothesis is supported, then dishes that contain the brunette hair dye will have the most growth of bacteria, showing that dark dye is most mutagenic. Dishes containing the blonde dye will have the least bacterial growth, and dishes containing the red dye will have an intermediate amount of growth.

Materials and Methods

The purpose of the materials and methods section is to give a detailed account of your experimental procedure. In order to be considered valid, the results of a scientific investigation must be able to be duplicated by other scientists working in other laboratories. It is therefore necessary to provide complete details of how your investigation was performed. In addition, scientists consult the Methods sections of published papers in order to learn techniques they can apply to their own work. So when you write this section, imagine that you are explaining what you did so that someone else can replicate your experiment exactly.

Use the following guidelines for the appropriate style in your Materials and Methods section.

- Use the past tense. Don't write as if you're giving instructions.

  INCORRECT: First you inoculate the agar plates with *Salmonella* bacteria.

  CORRECT: We inoculated the agar plates with *Salmonella* bacteria.

- Tell what you did in paragraph form. Don't write a recipe.

  INCORRECT:

  Step 1. Put 3 drops of hair dye on a filter paper.

  Step 2. Put the filter paper on the agar.

  CORRECT: We put 3 drops of hair dye on a filter paper, and then placed the filter paper on the agar.

- Be specific. Someone attempting to duplicate your experiment needs all the details.

  INCORRECT: We pipetted a sample of broth onto each plate.

  CORRECT: We pipetted a 1-mL sample of broth onto each plate.

Materials and Methods

Twelve petri dishes containing sterile agar growth medium that lacked histidine were prepared for us by the Biology Department prep staff. We used a broth that contained *Salmonella typhimurium* bacteria (strain his–) to "seed" the agar plates with bacteria. The *Salmonella* cultures were provided by the Microbiology Department. Each member of our group prepared four of the plates using the same technique. Using sterile pipets, we pipetted a 1-mL sample of the broth onto the agar in each plate. The broth was spread over the agar with a glass rod that had been sterilized by dipping the rod in alcohol and then burning off the alcohol in a flame. The rod was allowed to cool before being used to spread the broth. We learned these techniques from a lab manual (Sigmon, 2000).

All the significant details of how the experiment was done should be recounted here.

Following the standard technique for the Ames test (Sigmon, 2000), we used filter paper disks to apply the hair dye to the agar plates. We used 12 filter paper disks, each 1 cm in diameter. Using droppers, we put 1 drop of hair dye or sterile water (for a control) on each disk. Each treatment was thus replicated three times. The filter papers were air dried and then placed in the center of the petri dishes of agar. Forceps were always used to handle the filter paper.

Cite the source(s) of the techniques you used.

All the hair dyes used were Brand X. We chose Basic Brunette as the dark dye, Raspberry Red as the medium dye, and Bombshell Blonde as the light dye.

The reader should be informed about variables that must be standardized in order to have a successful experiment. If you later think of factors that should have been standardized but weren't, you should mention them in the Discussion section.

Two standardized variables in this experiment were temperature and the time that data were recorded. We put the petri dishes in an incubator set at 37°C and allowed them to incubate for 7 days. Each day at 11:30 A.M. a team member checked the plates and counted the number of bacterial colonies present. The results were recorded, tabulated, and distributed to the whole team.

Results

In the Results section you present the data in an organized, readable form. Numerical data are usually given in tables. Relationships between factors are often shown on graphs. Graphs, drawings, and anything else that is not a table is called a figure. Tables and figures should be numbered separately so that in the text you can refer to Table 1, Table 2, Figure 1, Figure 2, and so on. All tables and figures must have titles describing their contents.

Before beginning to write your Results section, you should review the data presentation section of Appendix A (Tools for Scientific Inquiry). Remember that you should not include raw data in your report. Also, results must be presented in some numerical fashion. A descriptive narrative is not acceptable.

Keep the following points in mind as you prepare your graphs:

- Use graph paper unless your graphs are computer generated.
- Label the axes completely.
- Use the entire area of the graph to display your data.
- Choose appropriate intervals and mark them evenly along the axes.
- If there is more than one set of data on the graph, be sure the reader can tell the lines apart. Include a legend (see Figure B.3).
- Each graph must have a descriptive title.

In addition to tables and figures, the Results section should include a brief paragraph that draws the reader's attention to the important pieces of data. However, you should save your explanations of *why* results are significant for the Discussion section.

Results

Author briefly points out important results that will be featured in the Discussion section.

Table B.1 shows that Basic Brunette had produced the greatest number of colonies at the end of 7 days. Figure B.3 shows that bacterial colonies on the dishes containing Raspberry Red and Bombshell Blonde only appeared on the last day, while colonies appeared on the Basic Brunette plates by Day 2 and steadily increased in number.

Table B.1
The total number of *Salmonella* his– bacterial colonies counted at the end of 7 days, using 3 replicates for each color of hair dye. The growth medium used was histidine–. Incubation was at 37°C.

Hair dye color	Number of colonies observed
Basic Brunette	36
Raspberry Red	4*
Bombshell Blonde	1
Water (control)	2

   * 3 of these colonies appeared to be contaminants.

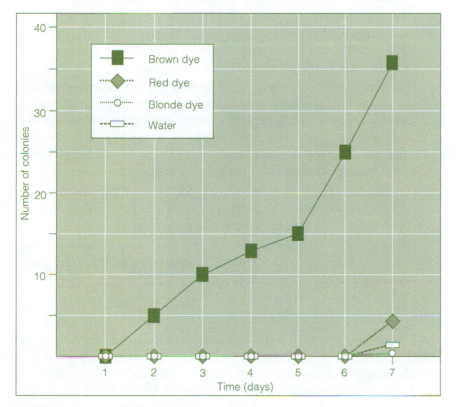

Figure B.3.
The total number of *Salmonella typhimurium* his– bacterial colonies (3 petri dishes per color) counted on each of 7 consecutive days. Incubation was at 37°C on growth medium that lacked histidine.

Discussion

In the Discussion section you should interpret the results, explain their significance, and discuss any weaknesses of the experimental methods or design. From the instructor's point of view, this is the most important section of your paper because it shows how well you understood your investigation. As you might expect, it is also the most difficult section to write.

You should complete the Introduction and Results sections before you begin writing the Discussion. Put them on the desk in front of you, along with your lab notebook open to your notes on the experiment, and begin writing your rough draft. You can outline the contents of the Discussion by taking the following steps.

1. Write down your hypothesis again. Look at the tables and/or figures you constructed for the Results section and determine whether you should accept or reject your hypothesis. (Did your experiment support your hypothesis or prove it false?)

2. Check the predictions you wrote in the Introduction section. Do your results confirm your predictions or not?

3. Write down the specific data (using the actual numbers) that led you to your conclusion about the hypothesis. If you have gotten additional results from other lab teams working on a similar problem, list that information also.

4. Write down what you know about the biology involved in your investigation. How do your results fit in with what you already know? Be sure to identify the sources of this information.

5. List any weaknesses you have identified in your experimental design. You must tell the reader how these weaknesses may have affected your results. Since your lab experiments are subject to limitations of time and facilities, you will not be able to do a "perfect" experiment. It is important for you to understand, and to acknowledge in your report, how these limitations affect the validity of your conclusions.

6. List any problems that arose during the experiment itself. Unforeseen difficulties with the procedures may have affected the data and should be described for the reader's consideration.

7. Review your experimental design and procedure. Consider how you might be able to get more specific or more reliable results by changing the experiment.

Discussion

The hypothesis is restated along with a brief statement of the investigator's conclusion.

Our experiment was designed to determine whether dark hair dye is more mutagenic to *Salmonella typhimurium* than light hair dye. Our results supported the hypothesis to the extent that Basic Brunette, a dark brown dye, was more mutagenic than Bombshell Blonde, a light hair dye. The results for Raspberry Red did not support the hypothesis.

Specific data are used in support of the conclusion. The author also explains what the data mean and compares the results to the predictions made before the experiment was performed.

As seen in Table B.1, the total number of colonies after 7 days of incubation (36) was greatest in the presence of Basic Brunette, a dark brown hair dye. This result means that the his– gene was mutated to his+ in 36 instances, supporting our hypothesis that the dark color is mutagenic. These results also confirm our prediction.

The plates containing the Bombshell Blonde hair dye and sterile water produced only one or two colonies each. Since the blonde dye did not produce any more mutations than water, which was the control, we can conclude that the blonde dye is not mutagenic. These results also support our hypothesis and prediction.

Raspberry Red hair dye produced four colonies, but three of these colonies, which were all on the same plate, were different colors from all the other colonies. Our lab instructor told us that these three colonies were contaminants, rather than *Salmonella typhimurium*, so only one of the colonies was actually mutated *Salmonella typhimurium* (Wright, personal communication). Red dye was therefore no more mutagenic than the control. This result was unexpected, since we had predicted that red dye would be intermediate in mutagenicity between brown and blonde.

The list of ingredients on the packages of hair dye gave some insight into the results for red dye. We had assumed that the different colors of dye would have the same pigments but in different amounts. However, Basic Brunette contains two pigments called chemicals A and B, Raspberry Red contains pigments called chemicals C and D, and Bombshell Blonde contains a pigment called chemical E. If we could test just these pigments using the Ames test, we could determine whether chemicals A and B are mutagenic.

For future experiments, we could also test different shades of Brand X hair color containing different pigments or different brands of hair color containing chemicals A and B.

The fact that the control plate produced two colonies suggests that some mutation takes place naturally. Starr and Taggart (1992) state that although some gene mutations are caused by mutagens, "other gene mutations are spontaneous; they are not induced by agents outside the cell." They further add that the rate of spontaneous mutations is relatively low. This supports the results from the plates containing water and Raspberry Red and Bombshell Blonde hair dyes.

Figure B.3 shows that five colonies had appeared in the Basic Brunette dishes by Day 2, and the number of colonies increased steadily over the entire 7 days of the experiment. The significance of this result is that the bacteria present were being mutated over time as the hair dye diffused to a wider and wider area of each dish. This means that the longer the bacteria are exposed to hair dye, the more mutations will occur. If we repeated this experiment, we would draw concentric rings around the filter paper disks and count the colonies appearing in each ring over time instead of just taking a total count of the dish.

The Ames test is not a definitive test for the cancer-causing ability of any chemical. It does indicate whether a chemical is mutagenic (can cause changes in the DNA). The Basic Brunette hair dye definitely showed signs of being mutagenic to *Salmonella* DNA, especially when there is longer exposure. Raspberry Red and Bombshell Blonde hair dyes gave no evidence of being mutagenic. Our results support the study cited in the *Greenville News* (2002), which found that some hair dyes are potential carcinogens.

Weaknesses in our experiment include the length of time allowed for incubation, the number of ingredients contained in the hair dyes, the fact that only one brand of dye was tested, and the contamination of one of the petri dishes. We could extend this experiment by testing the individual pigments in the hair dyes, using different brands of hair dye, testing the *Salmonella* his+ strain, and counting the colonies in concentric rings away from the hair dye disks.

The source of information obtained through conversation with a knowledgeable person should be cited as a "personal communication."

When the results are unexpected, you should try to explain why they differ from the prediction. Understanding the reasons for your results, whatever they are, is an important part of the scientific process.

The author uses this background information to show how her results fit in with what is already known on the subject.

Again, the author discusses specific data and interprets it using her knowledge of biological concepts. This leads to another suggestion for improving the experiment.

Author explains how the results of this investigation support published information.

Weaknesses of the experiment are pointed out. This shows that the author understood the method and results well enough to improve the experiment.

Conclusion

Your conclusions may be mentioned in three sections of your paper: Introduction, Discussion, and Conclusion, but they *must* be stated in the Conclusion. In the Conclusion section, try to rephrase your conclusion rather than repeat the exact wording used in a previous section. If your readers did not understand your initial version, another wording may clarify it.

The Conclusion should be brief (two or three sentences). It should repeat the significant results from your experiment, but should not contain any new information.

Conclusion

Basic Brunette caused 36 colonies of his– bacteria to grow on agar lacking histidine, so we concluded that this hair dye caused a mutation in the bacterial DNA. Raspberry Red and Bombshell Blonde dyes each produced a total of one mutated colony, so we concluded that these two dyes are not mutagenic.

Literature Cited

You must tell the reader exactly where to find the sources of information you used. In the text of the report, cite the source as (author, date) or author (date). For example:

(Author, date): A chemical that causes normal cells to become cancerous is called a carcinogen (Campbell, Reece, Mitchell, and Taylor, 2003).

Author (date): Campbell, Reece, Mitchell, and Taylor (2003) define a carcinogen as a chemical that causes normal cells to become cancerous.

At the end of the report, the Literature Cited section gives detailed information about the sources of information you used. The sources should be listed alphabetically by author. Any source that appears in your list must also be cited in the text of the report. The following examples illustrate one style. You may see slightly different citation styles used in different sources. Note that all citation styles include the same information; it is simply arranged differently.

Literature Cited

Atlas, C. *Microbiology*. New York: Macmillan Publishing Company, 1984.

Campbell, N., J. Reece, L. Mitchell, and M. Taylor. *Biology: Concepts and Connections*. San Francisco, CA: Benjamin/Cummings, 2003.

Greenville News. "New Study Links Hair Dye and Cancer." Greenville, SC: Greenville-Piedmont Publishing Company, June 4, 2002.

Sigmon, J. *Laboratory Techniques in Microbiology*. Clemson, SC: Clemson University, 2000.

Starr, C., and R. Taggart. *Biology: The Unity and Diversity of Life*, 6th ed. Belmont, CA: Wadsworth, 1992.

Wright, I. M., biology professor at Clemson University, personal communication, 2002.

Using Reference Materials Honestly

When your instructor reads your paper, he or she wants to evaluate your understanding of the biology involved in your experiment. It is essential for you to use your own words to explain your investigation rather than attempting to imitate your references. In addition, you could face a charge of plagiarism if "your" work is too similar to someone else's work.

Avoiding Plagiarism

Many students have difficulty trying to put information they've read into their own words. Typically, they change or rearrange a few words but leave the sentence essentially the same as it was written. For example, here is a sentence from a biology textbook (Campbell, Reece, Mitchell, and Taylor, 2003): "Cancer-causing agents, factors that alter DNA and make cells cancerous, are called carcinogens." A student changed this sentence to read, "Factors that cause cancer and alter DNA to make cells cancerous are called carcinogens." The student has tried to disguise this sentence from the textbook as his own work. Not only is the deception transparent to someone who has read the textbook, the awkwardness of the resulting sentence might even make the instructor wonder if the student knows what he is saying.

How can this problem be avoided?

- Take notes from your references, and then write from your notes rather than writing directly from the reference.

- Use more than one source of information for each topic so you won't get stuck on certain phrases.

- Think about what you know before you write it down. Digest and resynthesize the information you have read.

- Write down what you know quickly, without worrying about how well it's written. Later, revise your writing so a reader can understand it.

Using Quotations

Any sentence or phrase that is copied directly from a source must be placed in quotation marks.

The use of lengthy quotations indicates to the instructor that the student doesn't understand the subject well enough to explain it himself. For the lab reports you'll write for this course, it's recommended that direct quotations be no longer than one sentence.

Checklist for a Scientific Report

Title

_____Communicate the subject investigated in the paper.

Introduction

_____State the hypothesis.

_____Give well-defined reasons for making the hypothesis.

_____Explain the biological basis of the experiment.

_____Cite sources to substantiate background information.

_____Explain how the method used will produce information relevant to your hypothesis.

_____State a prediction based on your hypothesis. (If the hypothesis is supported, then the results will be . . .)

Materials and Methods

_____Use the appropriate style.

_____Give enough detail so the reader could duplicate your experiment.

_____State the control treatment, replication, and standardized variables that were used.

Results

_____Summarize the data (do not include raw data).

_____Present the data in an appropriate format (table or graph).

_____Present tables and figures neatly so they are easily read.

_____Label the axes of each graph completely.

_____Give units of measurement where appropriate.

_____Write a descriptive caption for each table and figure.

_____Include a short paragraph pointing out important results but do not interpret the data.

Discussion

_____State whether the hypothesis was supported or proven false by the results, or else state that the results were inconclusive.

_____Cite specific results that support your conclusions.

_____Give the reasoning for your conclusions.

_____Demonstrate that you understand the biological meaning of your results.

_____Compare the results with your predictions and explain any unexpected results.

_____Compare the results to other research or information available to you.

_____Discuss any weaknesses in your experimental design or problems with the execution of the experiment.

_____Discuss how you might extend or improve your experiment.

Conclusions

_____Restate your conclusion.

_____Restate important results.

Literature Cited

_____Use proper citation form in the text.

_____Use proper citation form in the Literature Cited section.

_____Refer in the text to any source listed in this section.

Acknowledgment

The sample report is based on a sample paper written by Janie Sigmon for students at Clemson University.

Credits and Trademarks

Photography Credits

Figures 4.1 and 4.9: Courtesy of Leica, Inc., Buffalo, New York 14240

Figure 8.1: Courtesy of University of Washington Department of Pathology.

Figure 9.1: Benjamin/Cummings.

Figures 11.1a, 11.1b, 11.1c: ©Janice Haney Carr/CDC

Figure 13.6: ©Harold W. Pratt/Biological Photo Service.

Figure 13.7: NIH/Photo Researchers, Inc.

Figure 13.12: ©Cath Ellis, Department of Zoology, University of Hull/ Science Photo Library/Photo Researchers, Inc.

Figure 15.4: National Library of Medicine.

Figure 19.1: ©M.W.F. Tweedie/Photo Researchers, Inc.

Page 16-23: CALVIN AND HOBBES copyright 1988 Watterson. Reprinted with permission of UNIVERSAL PRESS SYNDICATE. All rights reserved.

Color Insert

Plate 1 *Color charts:* Karl Miyajima.

Plate 2 *Liverwort:* ©Hal Horwitz/Corbis *Horsetails:* ©Zigrit/Fotolia *Ginko:* ©Pearson Education/Pearson Science *Cycad:* © Douglas Peebles/Corbis

Plate 3 *Sponges:* ©James Gritz/Getty Images, Inc.
Jellyfish: ©Paulo Curto/U.P./Bruce Coleman, Inc.
Coral: ©Linda Pitkin/Getty Images, Inc.
Sea anemone: ©Robert Yin/Corbis.

Plate 4 *Nudibranchs:* ©Robert Marien/Corbis *Squid:* ©Mexrix/Fotolia *Bivalve:* ©Kjell Sandved/Photo Researchers, Inc. *Conus shell:* ©Gator /Fotolia *Flatworm:* ©Stephen Fink/Corbis

Plate 5 *Shrimp:* ©Robert Yin/Corbis *Scorpion:* ©William Dow/Corbis *Brittle stars:* ©Jeffrey Rotman/Corbis
Wolf spider: ©Peter J. Bryant/Biological Photo Service.

Plate 6 *Sea star:* ©Steve Lovegrove/Fotolia. *Sea cucumber:* ©Hal Beral/Corbis. *Sea urchin:* ©Joze Maucec/Fotolia. *Snake:* ©Dorling Kindersley Media Library. *Sea squirt:* © Gary Bell/Getty Images, Inc.

Plate 7 *Beetle:* ©George D. Lepp/Corbis *Bee:* ©Becky Swora/Shutterstock *Fly:* ©Frank Lane Picture Agency/Corbis
Butterfly: ©Lee F. Snyder/Photo Researchers, Inc.

Plate 8 *Male corn plant:* ©Michael P. Gadomski/Photo Researchers, Inc. *Bat:* ©Merlin D. Tuttle/Bat Conservation International/Photo Researchers, Inc. *Female corn plant:* ©Shutterstock
Moth: ©Darlyne A. Murawski/Hulton Archive Photos/Getty Images Inc. *Hummingbird:* ©Michael & Patricia Fogden/Corbis

Text and Illustration Credits

The artists who contributed directly to *Laboratory Investigations for Biology* are listed on the copyright page at the front of the book. Illustrations from Campbell, Mitchell, and Reese, *Biology: Concepts and Connections* (©1994 The Benjamin/Cummings Publishing Company) that are used in *Laboratory Investigations for Biology* are the work of Nea Bisek, Mary Bryson, Barbara Cousins, Tom Dallman, Bill Glass, Illustrious, Inc., JAK Graphics, LTD, Georg Klatt, Laurie O'Keefe, Carla Simmons, Kevin Somerville, Terry Toyama, or Pamela Drury-Wattenmaker. Illustrations from Morgan and Carter, *Investigating Biology* (©1993 The Benjamin/Cummings Publishing Company) that are used in *Laboratory Investigations for Biology* are the work of Nea Bisek, Barbara Cousins, Terry Toyama, Valerie Felts, or John Norton.

The following figures are adapted from Judith Morgan and M. Eloise Carter, *Investigating Biology* (©1993 The Benjamin/Cummings Publishing Company): Figures 2.10, 2.11, 2.12, 2.13, 5.4, 5.6, 5.7, 6.8, 7.9, 9.5, 10.2, 11.4, 11.5, unnumbered 12.3, 12.12, 12.15, 13.1, unnumbered 13.3, 15.7, 17.2, unnumbered 17.4, 18.2, 18.3.

The following figures are adapted from Neil Campbell, Lawrence Mitchell, and Jane Reese, *Biology: Concepts and Connections* (Redwood City, CA: Benjamin/Cummings, 1994). Copyright ©1994 The Benjamin/Cummings Publishing Company): Figures 11.1, 15.1, 15.2, 16.1, 16.2, 16.5, 17.1, 18.1.

Figure 12.2 a & b was adapted from Neil Campbell, *Biology*, Third Edition (Redwood City, CA: Benjamin/Cummings, 1993). Copyright ©1993 The Benjamin/Cummings Publishing Company.

Preface Cartoon: Adapted from Jean Dickey, *Biology 105 Lab Manual (Clemson University)*, Contemporary Publishing. Copyright ©John Norton. Reprinted by permission of the artist.

Table 2.6: E.A. Pearce and G. Smith; adapted from *The Times Books World Weather Guide*, New York: Times Books, 1990.

Figures 9.5, 9.6, 9.7: Adapted from Carolina Biological Supply, *Wisconsin Fast Plants™ Manual*, (Burlington, NC: Carolina Biological Supply, 1989). Reproduced by permission of Carolina Biological Supply, Burlington, NC 27217 and Wisconsin Alumni Research Foundation, Madison, WI.

Figures 12.1, 12.10, 12.13: Adapted from Jean Dickey, *Biology 105 Lab Manual (Clemson University)*, Contemporary Publishing, pp. 158, 164, 184. Copyright ©Contemporary Publishing Company.

Trademarks

Alka-Seltzer is a registered trademark of Miles Inc.

Arm & Hammer is a registered trademark of Church & Dwight Co. Inc.

Ben-Gay is a registered trademark of Pfizer Inc.

Big Mac is a registered trademark of McDonald's Corporation.

Clorox is a registered trademark of The Clorox Company.

Dairy Ease, Lysol, and Milk of Magnesia are registered trademarks of Sterling Drug Inc.

Drano is a registered trademark of The Drackett Company.

Drierite is a registered trademark of W. A. Hammond Drierite Company.

Fast Plants is a trademark of J. M. Marketing Services, Inc.

The Food Processor is a trademark of ESHA Research Corporation.

Fruit-Fresh and Tums are registered trademarks of Beecham Inc.

Gatorade is a registered trademark of Stokely-Van Camp, Inc.

Jell-O is a registered trademark of General Foods Corporation.

Karo is a registered trademark of CPC International Inc.

Kool-Aid is a registered trademark of Perkins Products Company.

Lactaid is a registered trademark of Lactaid Inc.

Life Savers is a registered trademark of Life Savers, Inc.

Lime-A-Way is a registered trademark of Economics Laboratory, Inc.

Maalox is a registered trademark of Rorer Pharmaceutical Corporation.

Medeprin and Tylenol are trademarks of Johnson & Johnson.

Nutrasweet is a registered trademark of NutraSweet Company, Inc.

Parafilm is a registered trademark of Marathon Paper Mills Company.

Pepto-Bismol is a registered trademark of Norwich Eaton Pharmaceuticals, Inc.

PineSol is a registered trademark of American Cyanamid Company.

PROTOSLO is a registered trademark of Carolina Biological Supply Company.

Rolaids is a registered trademark of American Chicle Company.

7UP is a registered trademark of The Seven-Up Company.

Spectronic 20 is a registered trademark of Bausch & Lomb Incorporated.

Styrofoam is a registered trademark of The Dow Chemical Company.

Sweet'n Low is a registered trademark of Cumberland Packing Corp.

10-K is a registered trademark of Suntory Water Group, Inc.

Whatman is a registered trademark of Whatman Paper Limited.

Color chart for catecholase experiment

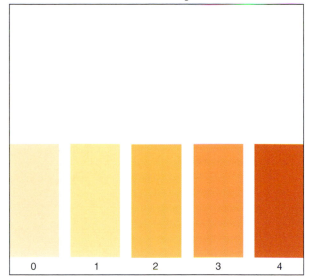

Compare your sample to this chart and record the number that most closely matches the intensity of your sample (Lab Topic 5).

Color chart for mitochondrial activity experiment

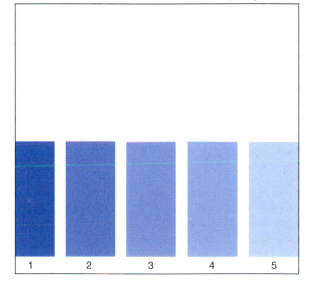

Compare your sample to this chart and record the number that most closely matches the intensity of your sample (Lab Topic 6).

Cycad plant, Hawaii (Lab Topic 12).

Giant Horsetail, *Equisetum teimateia*, Clark County, WA. (Lab Topic 12).

Fragrant Liverwort (Lab Topic 12).

Purple sponge (Lab Topic 13).

West Coast Sea Nettles, California (Lab Topic 13).

Gorgonian coral. Papua, New Guinea (Lab Topic 13).

Translucent tentacles extend from an orange sea anemone in the Philippines (Lab Topic 13).

Hilton's Aeolid, a nudibranch (Lab Topic 13).

Bivalve with foot and both siphons extended (Lab Topic 13).

Orange boring sponge and flatworm, California (Lab Topic 13).

Grass shrimp (Lab Topic 13).

Desert Hairy Scorpion (Lab Topic 13).

Wolf spider (Lab Topic 13).

A brittle star feeds on a blue sponge at night in the waters off the coast of Borneo in the Pacific Ocean. (Lab Topic 13).

Strongylocentrotus, **a sea urchin** (Lab Topic 13).

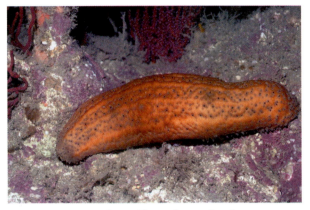

Warty sea cucumber off Catalina Island in California (Lab Topic 13).

Snake emerging from egg (Lab Topic 13).

Tunicates (Lab Topic 13).

Hover fly on white blossom (Lab Topic 18).

Striped beetle in center of yellow poppy (Lab Topic 18).

Gulf frittilary butterfly pollinating a *Pentas* flower. *Pentas* is in the madder family, which also includes gardenias and coffee plants (Lab Topic 18).

Bee pollinating a purple coneflower bloom.. Many flowers have petals that form a landing platform for pollinators (Lab Topic 18).

Greater short-nosed fruit bat pollinating a banana plant (Lab Topic 18).

Yucca moth on a yucca flower. The yucca moth exacts a price for pollinator service in this highly specialized relationship. The moth lays its eggs in the ovary of the yucca flower, and its developing larvae then feed on the yucca's seeds (Lab Topic 18).

A bronzy hermit hummingbird feeds on and pollinates a red passion flower in a cloud forest (Lab Topic 18).

Wind-pollinated flower: male and female corn plants (Lab Topic 18).